"The real cycle you're working on is a cycle called 'yourself.'"
ZEN AND THE ART OF MOTORCYCLE MAINTENANCE

"Profoundly important . . . full of insights into our most perplexing contemporary dilemmas . . . Intellectual entertainment of the highest order."
New York Times

"A miracle . . . [it] sparkles like an electric dream."
Village Voice

"[An] eloquent meditation on technology and philosophy."
Atlanta Journal-Constitution

"Part philosophical meditation and part travelogue . . . *Zen and the Art of Motorcycle Maintenance* . . . remains vital more than two decades later . . . Readers [are] drawn to the mesmerizing erudition of Pirsig's inquiry into science, religion, and humanism."
Albany Times-Union

"This multileveled story of a long bike trip across America by a man and his son has become a classic of philosophical literature."
Chicago Sun-Times

"It lodges in the mind . . . The book is inspired, original . . . The analogies with *Moby Dick* are patent. Robert Pirsig invites the prodigious comparison. What more can one say?"
The New Yorker

ZEN

AND
THE ART OF
MOTORCYCLE
MAINTENANCE

AN INQUIRY INTO VALUES

ROBERT M. PIRSIG

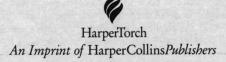

HarperTorch
An Imprint of HarperCollins*Publishers*

Special thanks should go to Stuart Cohen, who provided an office to write it in, and Mrs. Abigail Kenyon, for critical help with the early chapters.

❦

HARPERTORCH
An Imprint of HarperCollins*Publishers*
195 Broadway
New York, NY 10007

First HarperTorch paperback printing: May 2006
First HarperPerennial Modern Classics paperback printing: August 2005
First Quill paperback edition: May 1999

Printed in the United States of America

Visit HarperTorch on the World Wide Web at www.harpercollins.com

17 QGM 40 39 38 37 36 35 34

For my family

Author's Note

What follows is based on actual occurrences. Although much has been changed for rhetorical purposes, it must be regarded in its essence as fact. However, it should in no way be associated with that great body of factual information relating to orthodox Zen Buddhist practice. It's not very factual on motorcycles either.

And what is good, Phaedrus,
And what is not good—
Need we ask anyone to tell us these things?

Introduction

I suppose every writer dreams of the kind of success
Zen and the Art of Motorcycle Maintenance has had
since 1974, when it first appeared—rave reviews, millions of copies sold in twenty-seven languages, a description in the press as "the most widely read
philosophy book, ever."*

In the early seventies when it was being written, I
had those dreams, of course, but didn't let myself
dwell on them or express them publicly for fear they
would be interpreted as megalomania and a regression
to my former mental illness. Now the dreams are a reality and I don't have to worry about that anymore.

But rather than recount a success that everyone
knows, it would be of higher quality now to write
about the book's failures, and maybe help correct
them. There are two that stand out—a minor one and a
major one.

The minor one is that Phaedrus doesn't mean "wolf"
in Greek. That was a mistake that grew out of the actual experience at the University of Chicago in 1960
that appears in Part IV. The professor of philosophy

*The *London Telegraph* and BBC radio.

had mentioned that Plato liked to use names for his characters that suggested the nature of their personality, and in the dialogue *Phaedrus,* the likeness was made to a wolf. The professor, whose actual name I recall as either Lamm or Lamb, looked at me in such a way as to indicate he thought the title of wolf fit me. I was an outsider who seemed more interested in attacking what was being taught than learning from it. My hyperactive mind seized upon this as my definitive relationship to the school, and this worked its way into the book. But the character whom Plato likened to a wolf was not Phaedrus but Lycias, whose name is similar to the Greek *lykos* that does mean "wolf." As readers have pointed out to me many times, *Phaedrus* actually means "brilliant" or "radiant." I was lucky. It could just as easily have meant something much worse.

The second error is much more serious because it has obscured the fundamental meaning of the book. Many people have noticed that the ending somehow does not clear things up, that something is missing. Some have called it a "Hollywood ending" that undermines the artistic integrity of the book. They are right, but this is not because a Hollywood ending was intended. It is because a much different ending was intended that was not sufficiently clear. In the intended ending it is not the narrator who triumphs over a villainous Phaedrus. It is an honorable Phaedrus who triumphs over the narrator that has been maligning him all the time. This is now made clearer in this edition by using a sans-serif type for Phaedrus's voice.

To expand on this, let me go back to a creative writing seminar held on winter afternoons in the early 1950s at the University of Minnesota. The teacher was Allen Tate, a distinguished poet and literary critic. Our subject for many sessions was Henry James's *The Turn of the Screw,* in which a governess tries to shield her two protégés from a ghostly presence but in the end fails, and they are killed. I was completely convinced that this was just a straightforward ghost story, but Tate said no, Henry James is up to more than that. The governess is not the heroine of this story. She is the villainess. It is not the ghost who kills the children but the governess's hysterical belief that a ghost exists. I couldn't believe this at first, but reread the story and saw that Tate was right. You can interpret it either way.

How could I have missed it?

Tate explained that James was able to achieve this magic through the use of the first-person narrator. Tate said that the first person is the most difficult form because the writer is locked inside the head of the narrator and can't get out. He can't say "meanwhile, back at the ranch" as a transition to another subject because he is imprisoned forever inside the narrator. *But so is the reader!* And that is the strength of the first-person narrative. The reader does not see that the governess is the villainess because what the governess sees is all the reader ever sees.

Now come back to *Zen and the Art of Motorcycle Maintenance* and note the similarities. There is a narrator whose mind you never leave. He refers to an evil ghost named Phaedrus, but the only way you know this ghost is evil is because the narrator tells you so. During the story, Phaedrus appears in the narrator's dreams in such a way that you begin to see that not only is the narrator pursuing Phaedrus in order to destroy him but Phaedrus is also pursuing the narrator for the same purpose. Who will win?

There is a divided personality here: two minds fighting for the same body, a condition that inspired the original meaning of "schizophrenia." These two minds have different values as to what is important in life.

The narrator is primarily a person dominated by social values. As he says at the beginning, "I haven't really had a new idea in years." He never tells his story except in ways that are calculated to make you like him. His private thoughts he will share with you, but not with John or Sylvia or Chris or the DeWeeses. Above all, he does not want to be isolated from you—the reader—or from society around him. He maintains a careful position within the normal boundaries of his surrounding society because he has seen what has happened to Phaedrus who did not. He has learned his lesson. No more shock treatment for him. Only at one point does the narrator confess his secret: that he is a heretic who is congratulated by everyone for having saved his soul but who knows secretly that all he has saved is his skin.

There are only two others who know or sense this. Chris is one. He is going to pieces with confusion and grief as he looks for the father he remembers and loves and can't find anymore. Phaedrus is the other. He knows completely what the narrator is up to and despises him for it.

In Phaedrus's view the narrator is a sellout, a coward, who has abandoned truth for popularity and social acceptance by his psychiatrists, his family, his employers, and his social acquaintances. He sees that the narrator doesn't want to be honest anymore, just an accepted member of the community, bowing and accommodating his way through the rest of his years.

Phaedrus was dominated by intellectual values. He didn't give a damn who liked or didn't like him. He was single-mindedly pursuing a truth he felt was of staggering importance to the world. The world had no idea of what he was trying to do and it was trying to kill him for his trouble. Now he had been socially destroyed—silenced. But the residue of what he knew still lingered in the narrator's brain, and that was the source of the conflict.

In the end it is Chris's agony that releases Phaedrus. When Chris asks, "Were you really insane," and the answer is "No," it is not the narrator but Phaedrus who answers. And when Chris says, "I knew it," he also understands that for the first time on this whole trip he is talking to his long-lost father again. The tension is gone. They have won it. The dissembling narrator has vanished. "It's going to get better now," Phaedrus says. "You can sort of tell these things."

For more on the real Phaedrus, who is not a villainous ghost but rather a mild-mannered hyperintellectual, let me recommend *Lila,* a sequel that has been properly understood by very few. Let me also recommend www.moq.org on the Internet, a group that is among those few that understand it.

ZEN

AND

THE ART OF
MOTORCYCLE
MAINTENANCE

PART I

1

I can see by my watch, without taking my hand from the left grip of the cycle, that it is eight-thirty in the morning. The wind, even at sixty miles an hour, is warm and humid. When it's this hot and muggy at eight-thirty, I'm wondering what it's going to be like in the afternoon.

In the wind are pungent odors from the marshes by the road. We are in an area of the Central Plains filled with thousands of duck hunting sloughs, heading northwest from Minneapolis toward the Dakotas. This highway is an old concrete two-laner that hasn't had much traffic since a four-laner went in parallel to it several years ago. When we pass a marsh the air suddenly becomes cooler. Then, when we are past, it suddenly warms up again.

I'm happy to be riding back into this country. It is a kind of nowhere, famous for nothing at all and has an appeal because of just that. Tensions disappear along old roads like this. We bump along the beat-up concrete between the cattails and stretches of meadow and then more cattails and marsh grass. Here and there is a stretch of open water and if you look closely you can see wild ducks at the edge of the cattails. And turtles. . . . There's a red-winged blackbird.

I whack Chris's knee and point to it.

"What!" he hollers.

"Blackbird!"

He says something I don't hear. "What?" I holler back.

He grabs the back of my helmet and hollers up, "I've seen *lots* of those, Dad!"

"Oh!" I holler back. Then I nod. At age eleven you don't get very impressed with red-winged blackbirds.

You have to get older for that. For me this is all mixed with memories that he doesn't have. Cold mornings long ago when the marsh grass had turned brown and cattails were waving in the northwest wind. The pungent smell then was from muck stirred up by hip boots while we were getting in position for the sun to come up and the duck season to open. Or winters when the sloughs were frozen over and dead and I could walk across the ice and snow between the dead cattails and see nothing but grey skies and dead things and cold. The blackbirds were gone then. But now in July they're back and everything is at its alivest and every foot of these sloughs is humming and cricking and buzzing and chirping, a whole community of millions of living things living out their lives in a kind of benign continuum.

You see things vacationing on a motorcycle in a way that is completely different from any other. In a car you're always in a compartment, and because you're used to it you don't realize that through that car window everything you see is just more TV. You're a passive observer and it is all moving by you boringly in a frame.

On a cycle the frame is gone. You're completely in contact with it all. You're *in* the scene, not just watching it anymore, and the sense of presence is overwhelming. That concrete whizzing by five inches below your foot is the real thing, the same stuff you walk on, it's right there, so blurred you can't focus on it, yet you can put your foot down and touch it anytime, and the whole thing, the whole experience, is never removed from immediate consciousness.

Chris and I are traveling to Montana with some friends riding up ahead, and maybe headed farther than that. Plans are deliberately indefinite, more to travel than to arrive anywhere. We are just vacationing. Secondary roads are preferred. Paved county roads are the best, state highways are next. Freeways are the worst. We want to make good time, but for us now this is measured with emphasis on "good" rather than "time" and when you make that shift in emphasis the whole approach changes. Twisting hilly roads are long in terms of seconds but are much more enjoyable on a cycle where you bank into turns and don't get swung from side to side in any compartment. Roads with little traffic are more enjoyable, as well as safer. Roads free of drive-ins and billboards are better, roads where groves and meadows and orchards and lawns come almost to the shoulder, where kids wave to you when you ride by, where people look from their porches to see who it is, where when you stop to ask directions or information the answer tends to be longer than you want rather than short, where people ask where you're from and how long you've been riding.

It was some years ago that my wife and I and our friends first began to catch on to these roads. We took them once in a while for variety or for a shortcut to another main highway, and each time the scenery was grand and we left the road with a feeling of relaxation and enjoyment. We did this time after time before realizing what should have been obvious: these roads are truly different from the main ones. The whole pace of life and personality of the people who live along them are different. They're not going anywhere. They're not too busy to be courteous. The hereness and newness of things is something they know all about. It's the others, the ones who moved to the cities years ago and their lost offspring, who have all but forgotten it. The discovery was a real find.

I've wondered why it took us so long to catch on. We saw it and yet we didn't see it. Or rather we were trained *not* to see it. Conned, perhaps, into thinking that the real action was metropolitan and all this was just boring hinterland. It was a puzzling thing. The truth knocks on the door and you say, "Go away, I'm looking for the truth," and so it goes away. Puzzling.

But once we caught on, of course, nothing could keep us off these roads, weekends, evenings, vacations. We have become real secondary-road motorcycle buffs and found there are things you learn as you go.

We have learned how to spot the good ones on a map, for example. If the line wiggles, that's good. That means hills. If it appears to be the main route from a town to a city, that's bad. The best ones always connect nowhere with nowhere and have an alternate that gets you there quicker. If you are going northeast from a large town you never go straight out of town for any long distance. You go out and then start jogging north, then east, then north again, and soon you are on a secondary route that only the local people use.

The main skill is to keep from getting lost. Since the roads are used only by local people who know them by sight nobody complains if the junctions aren't posted. And often they aren't. When they are it's usually a small sign hiding unobtrusively in the weeds and that's all. County road-sign makers seldom tell you twice. If you miss that sign in the weeds that's *your* problem, not theirs. Moreover, you discover that the highway maps are often inaccurate about county roads. And from time to time you find your "county road" takes you onto a two-rutter and then a single rutter and then into a pasture and stops, or else it takes you into some farmer's backyard.

So we navigate mostly by dead reckoning, and deduction from what clues we find. I keep a compass in one pocket for overcast days when the sun doesn't show directions and have the map mounted in a special carrier on top of the gas tank where I can keep track of miles from the last junction and know what to look for. With those tools and a lack of pressure to "get somewhere" it works out fine and we just about have America all to ourselves.

On Labor Day and Memorial Day weekends we travel for miles on these roads without seeing another vehicle, then cross a federal highway and look at cars strung bumper to bumper to the horizon. Scowling faces inside. Kids crying in the back seat. I keep wishing there were some way to tell them something but they scowl and appear to be in a hurry, and there isn't. . . .

I have seen these marshes a thousand times, yet each time they're new. It's wrong to call them benign. You could just as well call them cruel and senseless, they are all of those things, but the *reality* of them over-whelms halfway conceptions. There! A huge flock of red-winged blackbirds ascends from nests in the cat-tails, startled by our sound. I swat Chris's knee a sec-ond time . . . then I remember he has seen them before.

"What?" he hollers again.

"Nothing."

"Well, *what*?"

"Just checking to see if you're still there," I holler, and nothing more is said.

Unless you're fond of hollering you don't make great conversations on a running cycle. Instead you spend your time being aware of things and meditating on them. On sights and sounds, on the mood of the weather and things remembered, on the machine and the countryside you're in, thinking about things at great leisure and length without being hurried and without feeling you're losing time.

What I would like to do is use the time that is com-ing now to talk about some things that have come to mind. We're in such a hurry most of the time we never get much chance to talk. The result is a kind of endless day-to-day shallowness, a monotony that leaves a per-son wondering years later where all the time went and sorry that it's all gone. Now that we do have some time, and know it, I would like to use the time to talk in some depth about things that seem important.

What is in mind is a sort of Chautauqua—that's the

only name I can think of for it—like the traveling tent-show Chautauquas that used to move across America, *this* America, the one that we are now in, an old-time series of popular talks intended to edify and entertain, improve the mind and bring culture and enlightenment to the ears and thoughts of the hearer. The Chautauquas were pushed aside by faster-paced radio, movies and TV, and it seems to me the change was not entirely an improvement. Perhaps because of these changes the stream of national consciousness moves faster now, and is broader, but it seems to run less deep. The old channels cannot contain it and in its search for new ones there seems to be growing havoc and destruction along its banks. In this Chautauqua I would like not to cut any new channels of consciousness but simply dig deeper into old ones that have become silted in with the debris of thoughts grown stale and platitudes too often repeated. "What's new?" is an interesting and broadening eternal question, but one which, if pursued exclusively, results only in an endless parade of trivia and fashion, the silt of tomorrow. I would like, instead, to be concerned with the question "What is best?", a question which cuts deeply rather than broadly, a question whose answers tend to move the silt downstream. There are eras of human history in which the channels of thought have been too deeply cut and no change was possible, and nothing new ever happened, and "best" was a matter of dogma, but that is not the situation now. Now the stream of our common consciousness seems to be obliterating its own banks, losing its central direction and purpose, flooding the lowlands, disconnecting and isolating the highlands and to no particular purpose other than the wasteful fulfillment of its own internal momentum. Some channel deepending seems called for.

Up ahead the other riders, John Sutherland and his wife, Sylvia, have pulled into a roadside picnic area. It's time to stretch. As I pull my machine beside them Sylvia is taking her helmet off and shaking her hair loose, while John puts his BMW up on the stand. Nothing is said. We have been on so many trips together we know from a glance how one another feels. Right now we are just quiet and looking around.

The picnic benches are abandoned at this hour of the morning. We have the whole place to ourselves. John goes across the grass to a cast-iron pump and starts pumping water to drink. Chris wanders down through some trees beyond a grassy knoll to a small stream. I am just staring around.

After a while Sylvia sits down on the wooden picnic bench and straightens out her legs, lifting one at a time slowly without looking up. Long silences mean gloom for her, and I comment on it. She looks up and then looks down again.

"It was all those people in the cars coming the other way," she says. "The first one looked so sad. And then the next one looked exactly the same way, and then the next one and the next one, they were all the same."

"They were just commuting to work."

She perceives well but there was nothing unnatural about it. "Well, you know, *work,*" I repeat. "Monday morning. Half asleep. Who goes to work Monday morning with a grin?"

"It's just that they looked so *lost,*" she says. "Like

they were all dead. Like a funeral procession." Then she puts both feet down and leaves them there.

I see what she is saying, but logically it doesn't go anywhere. You work to live and that's what they are doing. "I was watching swamps," I say.

After a while she looks up and says, "What did you see?"

"There was a whole flock of red-winged blackbirds. They rose up suddenly when we went by."

"Oh."

"I was happy to see them again. They tie things together, thoughts and such. You know?"

She thinks for a while and then, with the trees behind her a deep green, she smiles. She understands a peculiar language which has nothing to do with what you are saying. A daughter.

"Yes," she says. "They're beautiful."

"Watch for them," I say.

"All right."

John appears and checks the gear on the cycle. He adjusts some of the ropes and then opens the saddlebag and starts rummaging through. He sets some things on the ground. "If you ever need any rope, don't hesitate," he says. "God, I think I've got about *five* times what I need here."

"Not yet," I answer.

"Matches?" he says, still rummaging. "Sunburn lotion, combs, shoelaces . . . *shoelaces*? What do we need shoelaces for?"

"Let's not start *that*," Sylvia says. They look at each other deadpan and then both look over at me.

"Shoelaces can break anytime," I say solemnly. They smile, but not at each other.

Chris soon appears and it is time to go. While he gets ready and climbs on, they pull out and Sylvia waves. We are on the highway again, and I watch them gain distance up ahead.

The Chautauqua that is in mind for this trip was inspired by these two many months ago and perhaps, although I don't know, is related to a certain undercurrent of disharmony between them.

Disharmony I suppose is common enough in any marriage, but in their case it seems more tragic. To me, anyway.

It's not a personality clash between them; it's something else, for which neither is to blame, but for which neither has any solution, and for which I'm not sure I have any solution either, just ideas.

The ideas began with what seemed to be a minor difference of opinion between John and me on a matter of small importance: how much one should maintain one's own motorcycle. It seems natural and normal to me to make use of the small tool kits and instruction booklets supplied with each machine, and keep it tuned and adjusted myself. John demurs. He prefers to let a competent mechanic take care of these things so that they are done right. Neither viewpoint is unusual, and this minor difference would never have become magnified if we didn't spend so much time riding together and sitting in country roadhouses drinking beer and talking about whatever comes to mind. What comes to mind, usually, is whatever we've been thinking about in the half hour or forty-five minutes since we last talked to each other. When it's roads or weather or people or old memories or what's in the newspapers, the conversation just naturally builds pleasantly. But whenever the performance of the machine has been on my mind and gets into the conversation, the building stops. The conversation no longer moves forward. There is a silence and a break in the continuity. It is as though two old friends, a Catholic and Protestant, were sitting drinking beer, enjoying life, and the subject of birth control somehow came up. Big freeze-out.

And, of course, when you discover something like that it's like discovering a tooth with a missing filling. You can never leave it alone. You have to probe it, work around it, push on it, think about it, not because it's enjoyable but because it's on your mind and it won't get off your mind. And the more I probe and push on this subject of cycle maintenance the more irritated he gets, and of course that makes me want to probe and push all the more. Not deliberately to irritate him but because the irritation seems symptomatic of something deeper, something under the surface that isn't immediately apparent.

When you're talking birth control, what blocks it and freezes it out is that it's not a matter of more or fewer babies being argued. That's just on the surface. What's underneath is a conflict of faith, of faith in empirical social planning versus faith in the authority of God as revealed by the teachings of the Catholic Church. You can prove the practicality of planned parenthood till you get tired of listening to yourself and it's going to go nowhere because your antagonist isn't buying the assumption that anything socially practical is good per se. Goodness for him has other sources which he values as much as or more than social practicality.

So it is with John. I could preach the practical value and worth of motorcycle maintenance till I'm hoarse and it would make not a dent in him. After two sentences on the subject his eyes go completely glassy and he changes the conversation or just looks away. He doesn't want to hear about it.

Sylvia is completely with him on this one. In fact she is even more emphatic. "It's just a whole other thing," she says, when in a thoughtful mood. "Like garbage," she says, when not. They want *not* to understand it. Not to *hear* about it. And the more I try to fathom what makes me enjoy mechanical work and them hate it so, the more elusive it becomes. The ultimate cause of this originally minor difference of opinion appears to run way, way deep.

Inability on their part is ruled out immediately. They are both plenty bright enough. Either one of them could learn to tune a motorcycle in an hour and a half if they put their minds and energy to it, and the saving in money and worry and delay would repay them over and over again for their effort. And they *know* that. Or maybe they don't. I don't know. I never confront them with the question. It's better to just get along.

But I remember once, outside a bar in Savage, Minnesota, on a really scorching day when I just about let loose. We'd been in the bar for about an hour and we came out and the machines were so hot you could hardly get on them. I'm started and ready to go and there's John pumping away on the kick starter. I smell gas like we're next to a refinery and tell him so, thinking this is enough to let him know his engine's flooded.

"Yeah, I smell it too," he says and keeps on pumping. And he pumps and pumps and jumps and pumps and *I* don't know what more to say. Finally, he's really winded and sweat's running down all over his face and he can't pump anymore, and so I suggest taking out the plugs to dry them off and air out the cylinders while we go back for another beer.

Oh my God no! He doesn't want to get into all that stuff.

"All what stuff?"

"Oh, getting out the tools and all that stuff. There's no reason why it shouldn't start. It's a brand-new machine and I'm following the instructions perfectly. See, it's right on full choke like they say."

"Full *choke*!"

"That's what the instructions say."

"That's for when it's *cold*!"

"Well, we've been in there for a half an hour at least," he says.

It kind of shakes me up. "This is a hot day, John," I say. "And they take longer than that to cool off even on a freezing day."

He scratches his head. "Well, why don't they tell you that in the instructions?" He opens the choke and on the second kick it starts. "I guess that was it," he says cheerfully.

And the very next day we were out near the same area and it happened again. This time I was determined not to say a word, and when my wife urged me to go over and help him I shook my head. I told her that until he had a real felt need he was just going to resent help, so we went over and sat in the shade and waited.

I noticed he was being superpolite to Sylvia while he pumped away, meaning he was furious, and she was looking over with a kind of "Ye gods!" look. If he had asked any single question I would have been over in a second to diagnose it, but he wouldn't. It must have been fifteen minutes before he got it started.

Later we were drinking beer again over at Lake Minnetonka and everybody was talking around the table, but he was silent and I could see he was really tied up in knots inside. After all that time. Probably to get them untied he finally said, "You know . . . when it doesn't start like that it just . . . really turns me into a *monster* inside. I just get paranoic about it." This seemed to loosen him up, and he added, "They just had this *one* motorcycle, see? This *lemon.* And they didn't know what to do with it, whether to send it back to the factory or sell it for scrap or what . . . and then at the last moment they saw *me* coming. With eighteen hundred bucks in my pocket. And they knew their problems were over."

In a kind of singsong voice I repeated the plea for tuning and he tried hard to listen. He really tries hard sometimes. But then the block came again and he was off to the bar for another round for all of us and the subject was closed.

He is not stubborn, not narrow-minded, not lazy, not stupid. There was just no easy explanation. So it was left up in the air, a kind of mystery that one gives up on because there is no sense in just going round and round and round looking for an answer that's not there.

It occurred to me that maybe I was the odd one on the subject, but that was disposed of too. Most touring cyclists know how to keep their machines tuned. Car owners usually won't touch the engine, but every town of any size at all has a garage with expensive lifts, special tools and diagnostic equipment that the average owner can't afford. And a car engine is more complex and inaccessible than a cycle engine so there's more sense to this. But for John's cycle, a BMW R60, I'll bet there's not a mechanic between here and Salt Lake City. If his points or plugs burn out, he's done for. I *know* he doesn't have a set of spare points with him. He doesn't know what points are. If it quits on him in western South Dakota or Montana I don't know what he's going to do. Sell it to the Indians maybe. Right now I know what he's doing. He's carefully avoiding giving any thought whatsoever to the subject. The BMW is famous for not giving mechanical problems on the road and that's what he's counting on.

I might have thought this was just a peculiar attitude of theirs about motorcycles but discovered later that it extended to other things. . . . Waiting for them to get going one morning in their kitchen I noticed the sink faucet was dripping and remembered that it was dripping the last time I was there before and that in fact it had been dripping as long as I could remember. I commented on it and John said he had tried to fix it with a new faucet washer but it hadn't worked. That was all he said. The presumption left was that that was the end of the matter. If you try to fix a faucet and your fixing doesn't work then it's just your lot to live with a dripping faucet.

This made me wonder to myself if it got on their nerves, this drip-drip-drip, week in, week out, year in, year out, but I could not notice any irritation or concern about it on their part, and so concluded they just aren't bothered by things like dripping faucets. Some people aren't.

What it was that changed this conclusion, I don't remember . . . some intuition, some insight one day, perhaps it was a subtle change in Sylvia's mood whenever the dripping was particularly loud and she was trying to talk. She has a very soft voice. And one day when she was trying to talk above the dripping and the kids came in and interrupted her she lost her temper at them. It seemed that her anger at the kids would not have been nearly as great if the faucet hadn't also been dripping when she was trying to talk. It was the combined dripping and loud kids that blew her up. What struck me hard then was that she was *not* blaming the faucet, and that she was *deliberately* not blaming the faucet. She wasn't ignoring that faucet at all! She was *suppressing* anger at that faucet and that goddamned dripping faucet was just about *killing* her! But she could not admit the importance of this for some reason.

Why suppress anger at a dripping faucet? I wondered.

Then that patched in with the motorcycle maintenance and one of those light bulbs went on over my head and I thought, Ahhhhhhhh!

It's not the motorcycle maintenance, not the faucet. It's all of technology they can't take. And then all sorts of things started tumbling into place and I knew that was it. Sylvia's irritation at a friend who thought computer programming was "creative." All their drawings and paintings and photographs without a technological thing in them. Of course she's not going to get mad at that faucet, I thought. You always suppress momentary anger at something you deeply and permanently hate. Of course John signs off every time the subject of cycle repair comes up, even when it is obvious he is suffering for it. That's technology. And sure, of course, obviously. It's so simple when you see it. To get away from technology out into the country in the fresh air and sunshine is why they are on the motorcycle in the first place. For me to bring it back to them just at the point and place where they think they have finally escaped it just frosts both of them, tremendously. That's why the conversation always breaks and freezes when the subject comes up.

Other things fit in too. They talk once in a while in as few pained words as possible about "it" or "it all" as in the sentence, "There is just no escape from it." And if I asked, "From what?" the answer might be "The whole thing," or "The whole organized bit," or even "The system." Sylvia once said defensively, "Well, *you* know how to *cope* with it," which puffed me up so much at the time I was embarrassed to ask what "it" was and so remained somewhat puzzled. I thought it was something more mysterious than technology. But now I see that the "it" was mainly, if not entirely, technology. But, that doesn't sound right either. The "it" is a kind of force that gives rise to technology, something undefined, but inhuman, mechanical, lifeless, a blind monster, a death force. Something hideous they are running from but know they can never escape. I'm putting it way too heavily here but in a less emphatic and less defined way this is what it is. Somewhere there are people who understand it and run it but those are technologists, and they speak an inhuman language when describing what they do. It's all parts and relationships of unheard-of things that never make any sense no matter how often you hear about them. And their things, their monster keeps eating up land and polluting their air and lakes, and there is no way to strike back at it, and hardly any way to escape it.

That attitude is not hard to come to. You go through a heavy industrial area of a large city and there it all is, the technology. In front of it are high barbed-wire fences, locked gates, signs saying NO TRESPASSING, and beyond, through sooty air, you see ugly strange shapes of metal and brick whose purpose is unknown, and whose masters you will never see. What it's for you don't know, and why it's there, there's no one to tell, and so all you can feel is alienated, estranged, as though you didn't belong there. Who owns and understands this doesn't want you around. All this technology has somehow made you a stranger in your own land. Its very shape and appearance and mysteriousness say, "Get out." You know there's an explanation for all this somewhere and what it's doing undoubtedly serves mankind in some indirect way but that isn't what you see. What you see is the NO TRESPASSING, KEEP OUT signs and not anything serving people but little people, like ants, serving these strange, incomprehensible shapes. And you think, even if I were a part of this, even if I were not a stranger, I would be just another ant serving the shapes. So the final feeling is hostile, and I think that's ultimately what's involved with this otherwise unexplainable attitude of John and Sylvia. Anything to do with valves and shafts and wrenches is a part of *that* dehumanized world, and they would rather not think about it. They don't want to get into it.

If this is so, they are not alone. There is no question that they have been following their natural feelings in this and not trying to imitate anyone. But many others are also following their natural feelings and not trying to imitate anyone and the natural feelings of very many people are similar on this matter; so that when you look at them collectively, as journalists do, you get the illusion of a mass movement, an antitechnological mass movement, an entire political antitechnological left emerging, looming up from apparently nowhere, saying, "Stop the technology. Have it somewhere else. Don't have it here." It is still restrained by a thin web of logic that points out that without the factories there are no jobs or standard of living. But there are human forces stronger than logic. There always have been, and if they become strong enough in their hatred of technology that web can break.

Clichés and stereotypes such as "beatnik" or "hippie" have been invented for the antitechnologists, the antisystem people, and will continue to be. But one does not convert individuals into mass people with the simple coining of a mass term. John and Sylvia are not mass people and neither are most of the others going their way. It is against being a mass person that they seem to be revolting. And they feel that technology has got a lot to do with the forces that are trying to turn them into mass people and they don't like it. So far it's still mostly a passive resistance, flights into the rural areas when they are possible and things like that, but it doesn't always have to be this passive.

I disagree with them about cycle maintenance, but not because I am out of sympathy with their feelings about technology. I just think that their flight from and hatred of technology is self-defeating. The Buddha, the Godhead, resides quite as comfortably in the circuits of a digital computer or the gears of a cycle transmission as he does at the top of a mountain or in the petals of a flower. To think otherwise is to demean the Buddha—which is to demean oneself. That is what I want to talk about in this Chautauqua.

We're out of the marshes now, but the air is still so humid you can look straight up directly at the yellow circle of the sun as if there were smoke or smog in the sky. But we're in the green countryside now. The farmhouses are clean and white and fresh. And there's no smoke or smog.

2

The road winds on and on . . . we stop for rests and lunch, exchange small talk, and settle down to the long ride. The beginning fatigue of afternoon balances the excitement of the first day and we move steadily, not fast, not slow.

We have picked up a southwest side wind, and the cycle cants into the gusts, seemingly by itself, to counter their effect. Lately there's been a sense of something peculiar about this road, apprehension about something, as if we were being watched or followed. But there is not a car anywhere ahead, and in the mirror are only John and Sylvia way behind.

We are not in the Dakotas yet, but the broad fields show we are getting nearer. Some of them are blue with flax blossoms moving in long waves like the surface of the ocean. The sweep of the hills is greater than before and they now dominate everything else, except the sky, which seems wider. Farmhouses in the distance are so small we can hardly see them. The land is beginning to open up.

There is no one place or sharp line where the Central Plains end and the Great Plains begin. It's a gradual change like this that catches you unawares, as if you were sailing out from a choppy coastal harbor, noticed that the waves had taken on a deep swell, and turned back to see that you were out of sight of land. There are fewer trees here and suddenly I am aware they are no longer native. They have been brought here and planted around houses and between fields in rows to break up the wind. But where they haven't been planted there is no underbrush, no second-growth saplings—only grass, sometimes with wildflowers and weeds, but mostly grass. This is grassland now. We are on the prairie.

I have a feeling none of us fully understands what four days on this prairie in July will be like. Memories of car trips across them are always of flatness and great emptiness as far as you can see, extreme monotony and boredom as you drive for hour after hour, getting nowhere, wondering how long this is going to last without a turn in the road, without a change in the land going on and on to the horizon.

John was worried Sylvia would not be up to the discomfort of this and planned to have her fly to Billings, Montana, but Sylvia and I both talked him out of it. I argued that physical discomfort is important only when the mood is wrong. Then you fasten on to whatever thing is uncomfortable and call that the cause. But if the mood is right, then physical discomfort doesn't mean much. And when thinking about Sylvia's moods and feelings, I couldn't see her complaining.

Also, to arrive in the Rocky Mountains by plane would be to see them in one kind of context, as pretty scenery. But to arrive after days of hard travel across the prairies would be to see them in another way, as a goal, a promised land. If John and I and Chris arrived with this feeling and Sylvia arrived seeing them as "nice" and "pretty," there would be more disharmony among us than we would get from the heat and monotony of the Dakotas. Anyway, I like to talk to her and I'm thinking of myself too.

In my mind, when I look at these fields, I say to her, "See? . . . See?" and I think she does. I hope later she will see and feel a thing about these prairies I have given up talking to others about; a thing that exists here because everything else does not and can be noticed because other things are absent. She seems so depressed sometimes by the monotony and boredom of her city life, I thought maybe in this endless grass and wind she would see a thing that sometimes comes when monotony and boredom are accepted. It's here, but I have no names for it.

Now on the horizon I see something else I don't think the others see. Far off to the southwest—you can see it only from the top of this hill—the sky has a dark edge. Storm coming. That may be what has been bothering me. Deliberately shutting it out of mind, but knowing all along that with this humidity and wind it was more than likely. It's too bad, on the first day, but as I said before, on a cycle you're *in* the scene, not just watching it, and storms are definitely part of it.

If it's just thunderheads or broken line squalls you can try to ride around them, but this one isn't. That long dark streak without any preceding cirrus clouds is a cold front. Cold fronts are violent and when they are from the southwest, they are the most violent. Often they contain tornadoes. When they come it's best to just hole up and let them pass over. They don't last long and the cool air behind them makes good riding.

Warm fronts are the worst. They can last for days. I remember Chris and I were on a trip to Canada a few years ago, got about 130 miles and were caught in a warm front of which we had plenty of warning but which we didn't understand. The whole experience was kind of dumb and sad.

We were on a little six-and-one-half-horsepower cycle, way overloaded with luggage and way underloaded with common sense. The machine could do only about forty-five miles per hour wide open against a moderate head wind. It was no touring bike. We reached a large lake in the North Woods the first night and tented amid rainstorms that lasted all night long. I forgot to dig a trench around the tent and at about two in the morning a stream of water came in and soaked both sleeping bags. The next morning we were soggy and depressed and hadn't had much sleep, but I thought that if we just got riding the rain would let up after a while. No such luck. By ten o'clock the sky was so dark all the cars had their headlights on. And then it really came down.

We were wearing the ponchos which had served as a tent the night before. Now they spread out like sails and slowed our speed to thirty miles an hour wide open. The water on the road became two inches deep. Lightning bolts came crashing down all around us. I remember a woman's face looking astonished at us from the window of a passing car, wondering what in earth we were doing on a motorcycle in this weather. I'm sure I couldn't have told her.

The cycle slowed down to twenty-five, then twenty. Then it started missing, coughing and popping and sputtering until, barely moving at five or six miles an hour, we found an old run-down filling station by some cutover timberland and pulled in.

At the time, like John, I hadn't bothered to learn much about motorcycle maintenance. I remember holding my poncho over my head to keep the rain from the tank and rocking the cycle between my legs. Gas seemed to be sloshing around inside. I looked at the plugs, and looked at the points, and looked at the carburetor, and pumped the kick starter until I was exhausted.

We went into the filling station, which was also a combination beer joint and restaurant, and had a meal of burned-up steak. Then I went back out and tried it again. Chris kept asking questions that started to anger me because he didn't see how serious it was. Finally I saw it was no use, gave it up, and my anger at him disappeared. I explained to him as carefully as I could that it was all over. We weren't going anywhere by cycle on this vacation. Chris suggested things to do like check the gas, which I had done, and find a mechanic. But there weren't any mechanics. Just cutover pine trees and brush and rain.

I sat in the grass with him at the shoulder of the road, defeated, staring into the trees and underbrush. I answered all of Chris's questions patiently and in time they became fewer and fewer. And then Chris finally understood that our cycle trip was really over and began to cry. He was eight then, I think.

We hitchhiked back to our own city and rented a trailer and put it on our car and came up and got the cycle, and hauled it back to our own city and then started out all over again by car. But it wasn't the same. And we didn't really enjoy ourselves much.

Two weeks after the vacation was over, one evening after work, I removed the carburetor to see what was wrong but still couldn't find anything. To clean off the grease before replacing it, I turned the stopcock on the tank for a little gas. Nothing came out. The tank was out of gas. I couldn't believe it. I can still hardly believe it.

I have kicked myself mentally a hundred times for that stupidity and don't think I'll ever really, finally get over it. Evidently what I saw sloshing around was gas in the reserve tank which I had never turned on. I didn't check it carefully because I assumed the rain had caused the engine failure. I didn't understand then how foolish quick assumptions like that are. Now we are on a twenty-eight-horse machine and I take the maintenance of it very seriously.

All of a sudden John passes me, his palm down, signaling a stop. We slow down and look for a place to pull off on the gravelly shoulder. The edge of the concrete is sharp and the gravel is loose and I'm not a bit fond of his maneuver.

Chris asks, "What are we stopping for?"

"I think we missed our turn back there," John says.

I look back and see nothing. "I didn't see any sign," I say.

John shakes his head. "Big as a barn door."

"Really?"

He and Sylvia both nod.

He leans over, studies my map and points to where the turn was and then to a freeway overpass beyond it. "We've already crossed this freeway," he says. I see he is right. Embarrassing. "Go back or go ahead?" I ask.

He thinks about it. "Well, I guess there's really no reason to go back. All right. Let's just go ahead. We'll get there one way or another."

And now tagging along behind them I think, Why should I do a thing like that? I hardly noticed the freeway. And just now I forgot to tell them about the storm. Things are getting a little unsettling.

The storm cloud bank is larger now but it is not moving in as fast as I thought it would. That's not so good. When they come in fast they leave fast. When they come in slow like this you can get stuck for quite a time.

I remove a glove with my teeth, reach down and feel the aluminum side cover of the engine. The temperature is fine. Too warm to leave my hand there, not so hot I get a burn. Nothing wrong there.

On an air-cooled engine like this, extreme overheating can cause a "seizure." This machine has had one . . . in fact, three of them. I check it from time to time the same way I would check a patient who has had a heart attack, even though it seems cured.

In a seizure, the pistons expand from too much heat, become too big for the walls of the cylinders, seize them, melt to them sometimes, and lock the engine and rear wheel and start the whole cycle into a skid. The first time this one seized, my head was pitched over the front wheel and my passenger was almost on top of me. At about thirty it freed up again and started to run but I pulled off the road and stopped to see what was wrong. All my passenger could think to say was "What did you do *that* for?"

I shrugged and was as puzzled as he was, and stood there with the cars whizzing by, just staring. The engine was so hot the air around it shimmered and we could feel the heat radiate. When I put a wet finger on it, it sizzled like a hot iron and we rode home, slowly, with a new sound, a slap that meant the pistons no longer fit and an overhaul was needed.

I took this machine into a shop because I thought it wasn't important enough to justify getting into myself, having to learn all the complicated details and maybe having to order parts and special tools and all that time-dragging stuff when I could get someone else to do it in less time—sort of John's attitude.

The shop was a different scene from the ones I remembered. The mechanics, who had once all seemed like ancient veterans, now looked like children. A radio was going full blast and they were clowning around and talking and seemed not to notice me. When one of them finally came over he barely listened to the piston slap before saying, "Oh yeah. Tappets."

Tappets? I should have known then what was coming.

Two weeks later I paid their bill for 140 dollars, rode the cycle carefully at varying low speeds to wear it in and then after one thousand miles opened it up. At about seventy-five it seized again and freed at thirty, the same as before. When I brought it back they accused me of not breaking it in properly, but after much argument agreed to look into it. They overhauled it again and this time took it out themselves for a high-speed road test.

It seized on *them* this time.

After the third overhaul two months later they replaced the cylinders, put in oversize main carburetor jets, retarded the timing to make it run as coolly as possible and told me, "Don't run it fast."

It was covered with grease and did not start. I found the plugs were disconnected, connected them and started it, and now there really *was* a tappet noise. They hadn't adjusted them. I pointed this out and the kid came with an open-end adjustable wrench, set wrong, and swiftly rounded both of the sheet-aluminum tappet covers, ruining both of them.

"I hope we've got some more of those in stock," he said.

I nodded.

He brought out a hammer and cold chisel and started to pound them loose. The chisel punched through the aluminum cover and I could see he was pounding the chisel right into the engine head. On the next blow he missed the chisel completely and struck the head with the hammer, breaking off a portion of two of the cooling fins.

"Just stop," I said politely, feeling this was a bad dream. "Just give me some new covers and I'll take it the way it is."

I got out of there as fast as possible, noisy tappets, shot tappet covers, greasy machine, down the road, and then felt a bad vibration at speeds over twenty. At the curb I discovered two of the four engine-mounting bolts were missing and a nut was missing from the third. The whole engine was hanging on by only one bolt. The overhead-cam chain-tensioner bolt was also missing, meaning it would have been hopeless to try to adjust the tappets anyway. Nightmare.

The thought of John putting his BMW into the hands of one of those people is something I have never brought up with him. Maybe I should.

I found the cause of the seizures a few weeks later, waiting to happen again. It was a little twenty-five-cent pin in the internal oil-delivery system that had been sheared and was preventing oil from reaching the head at high speeds.

The question *why* comes back again and again and has become a major reason for wanting to deliver this Chautauqua. Why did they butcher it so? These were not people running away from technology, like John and Sylvia. These were the technologists themselves. They sat down to do a job and they performed it like chimpanzees. Nothing personal in it. There was no obvious reason for it. And I tried to think back into that shop, that nightmare place, to try to remember anything that could have been the cause.

The radio was a clue. You can't really think hard about what you're doing and listen to the radio at the same time. Maybe they didn't see their job as having anything to do with hard thought, just wrench twiddling. If you can twiddle wrenches while listening to the radio that's more enjoyable.

Their speed was another clue. They were really slopping things around in a hurry and not looking where they slopped them. More money that way—if you don't stop to think that it usually takes longer or comes out worse.

But the biggest clue seemed to be their expressions. They were hard to explain. Good-natured, friendly, easygoing—and uninvolved. They were like spectators. You had the feeling they had just wandered in there themselves and somebody had handed them a wrench. There was no identification with the job. No saying, "I am a mechanic." At 5 P.M. or whenever their eight hours were in, you knew they would cut it off and not have another thought about their work. They were already trying not to have any thoughts about their work *on* the job. In their own way they were achieving the same thing John and Sylvia were, living with technology without really having anything to do with it. Or rather, they had something to do with it, but their own selves were outside of it, detached, removed. They were involved in it but not in such a way as to care.

Not only did these mechanics not find that sheared pin, but it was clearly a mechanic who had sheared it in the first place, by assembling the side cover plate improperly. I remembered the previous owner had said a mechanic had told him the plate was hard to get on. That was why. The shop manual had warned about this, but like the others he was probably in too much of a hurry or he didn't care.

While at work I was thinking about this same lack of care in the digital computer manuals I was editing. Writing and editing technical manuals is what I do for a living the other eleven months of the year and I knew they were full of errors, ambiguities, omissions and information so completely screwed up you had to read them six times to make any sense out of them. But what struck me for the first time was the agreement of these manuals with the spectator attitude I had seen in the shop. These were spectator manuals. It was built into the format of them. Implicit in every line is the idea that "Here is the machine, isolated in time and in space from everything else in the universe. It has no relationship to you, you have no relationship to it, other than to turn certain switches, maintain voltage levels, check for error condition . . ." and so on. That's it. The mechanics in their attitude toward the machine were really taking no different attitude from the manual's toward the machine, or from the attitude I had when I brought it in there. We were all spectators. And it occurred to me there *is* no manual that deals with the *real* business of motorcycle maintenance, the most important aspect of all. Caring about what you are doing is considered either unimportant or taken for granted.

On this trip I think we should notice it, explore it a little, to see if in that strange separation of what man is from what man does we may have some clues as to what the hell has gone wrong in this twentieth century. I don't want to hurry it. That itself is a poisonous twentieth-century attitude. When you want to hurry something, that means you no longer care about it and want to get on to other things. I just want to get at it slowly, but carefully and thoroughly, with the same attitude I remember was present just before I found that sheared pin. It was that attitude that found it, nothing else.

I suddenly notice the land here has flattened into a Euclidian plane. Not a hill, not a bump anywhere. This means we have entererd the Red River Valley. We will soon be into the Dakotas.

3

By the time we are out of the Red River Valley the storm clouds are everywhere and almost upon us.

John and I have discussed the situation in Breckenridge and decided to keep going until we have to stop.

That shouldn't be long now. The sun is gone, the wind is blowing cold, and a wall of differing shades of grey looms around us.

It seems huge, overpowering. The prairie here is huge but above it the hugeness of this ominous grey mass ready to descend is frightening. We are traveling at its mercy now. When and where it will come is nothing we can control. All we can do is watch it move in closer and closer.

Where the darkest grey has come down to the ground, a town that was seen earlier, some small buildings and a water tower, has disappeared. It will be on us soon now. I don't see any towns ahead and we are just going to have to run for it.

I pull up alongside John and throw my hand ahead

in a "Speed up!" gesture. He nods and opens up. I let him get ahead a little, then pick up to his speed. The engine responds beautifully—seventy . . . eighty . . . eighty-five . . . we are really feeling the wind now and I drop my head to cut down the resistance . . . ninety. The speedometer needle swings back and forth but the tach reads a steady nine thousand . . . about ninety-five miles an hour . . . and we hold this speed . . . moving. Too fast to focus on the shoulder of the road now . . . I reach forward and flip the headlight switch just for safety. But it is needed anyway. It is getting very dark.

We whizz through the flat open land, not a car anywhere, hardly a tree, but the road is smooth and clean and the engine now has a "packed," high rpm sound that says it's right on. It gets darker and darker.

A flash and *Ka-wham!* of thunder, one right on top of the other. That shook me, and Chris has got his head against my back now. A few warning drops of rain . . . at this speed they are like needles. A second flash-*WHAM* and everything brilliant . . . and then in the brilliance of the next flash that farmhouse . . . that windmill . . . oh, my God, he's *been* here! . . . throttle off . . . this is *his* road . . . a fence and trees . . . and the speed drops to seventy, then sixty, then fifty-five and I hold it there.

"Why are we slowing down?" Chris shouts.

"Too fast!"

"No, it isn't!"

I nod yes.

The house and water tower have gone by and then a small drainage ditch appears and a crossroad leading off to the horizon. Yes . . . that's right, I think. That's exactly right.

"They're way ahead of us!" Chris hollers. "Speed up!"

I turn my head from side to side.

"Why not?" he hollers.

"Not safe!"

"They're gone!"

"They'll wait!"

"Speed up!"

"No." I shake my head. It's just a feeling. On a cycle you trust them and we stay at fifty-five.

The first rain begins now but up ahead I see the lights of a town . . . I knew it would be there.

When we arrive John and Sylvia are there under the first tree by the road, waiting for us.

"What happened to you?"

"Slowed down."

"Well, we know *that*. Something wrong?"

"No. Let's get out of this rain."

John says there is a motel at the other end of town, but I tell him there's a better one if you turn right, at a row of cottonwoods a few blocks down.

We turn at the cottonwoods and travel a few blocks, and a small motel appears. Inside the office John looks around and says, "This *is* a good place. When were you here before?"

"I don't remember," I say.

"Then how did you know about this?"

"Intuition."

He looks at Sylvia and shakes his head.

Sylvia has been watching me silently for some time. She notices my hands are unsteady as I sign in. "You look awfully pale," she says. "Did that lightning shake you up?"

"No."

"You look like you'd seen a ghost."

John and Chris look at me and I turn away from them to the door. It is still raining hard, but we make a run for it to the rooms. The gear on the cycles is protected and we wait until the storm passes over before removing it.

After the rain stops, the sky lightens a little. But from the motel courtyard, I see past the cottonwoods that a second darkness, that of night, is about to come on. We walk into town, have supper, and by the time we get back, the fatigue of the day is really on me. We rest, almost motionless, in the metal armchairs of the motel courtyard, slowly working down a pint of whiskey that John brought with some mix from the motel cooler. It goes down slowly and agreeably. A cool night wind rattles the leaves of the cottonwoods along the road.

Chris wonders what we should do next. Nothing tires this kid. The newness and strangeness of the motel surroundings excite him and he wants us to sing songs as they did at camp.

"We're not very good at songs," John says.

"Let's tell stories then," Chris says. He thinks for a while. "Do you know any good ghost stories? All the kids in our cabin used to tell ghost stories at night."

"You tell *us* some," John says.

And he does. They are kind of fun to hear. Some of them I haven't heard since I was his age. I tell him so, and Chris wants to hear some of mine, but I can't remember any.

After a while he says, "Do you believe in ghosts?"

"No," I say.

"Why not?"

"Because they are *un*-sci-en-*ti*-fic."

The way I say this makes John smile. "They contain no matter," I continue, "and have no energy and therefore, according to the laws of science, do not exist except in people's minds."

The whiskey, the fatigue and the wind in the trees start mixing in my mind. "Of course," I add, "the laws of science contain no matter and have no energy either and therefore do not exist except in people's minds. It's best to be completely scientific about the whole thing and refuse to believe in either ghosts or the laws of science. That way you're safe. That doesn't leave you very much to believe in, but that's scientific too."

"I don't know what you're talking about," Chris says.

"I'm being kind of facetious."

Chris gets frustrated when I talk like this, but I don't think it hurts him.

"One of the kids at YMCA camp says he believes in ghosts."

"He was just spoofing you."

"No, he wasn't. He said that when people haven't been buried right, their ghosts come back to haunt people. He really believes in that."

"He was just spoofing you," I repeat.

"What's his name?" Sylvia says.

"Tom White Bear."

John and I exchange looks, suddenly recognizing the same thing.

"Ohhh, *Indian*!" he says.

I laugh. "I guess I'm going to have to take that back a little," I say. "I was thinking of European ghosts."

"What's the difference?"

John roars with laughter. "He's got you," he says.

I think a little and say, "Well, Indians sometimes have a different way of looking at things, which I'm not saying is completely wrong. Science isn't part of the Indian tradition."

"Tom White Bear said his mother and dad told him not to believe all that stuff. But he said his grandmother whispered it was true anyway, so he believes it."

He looks at me pleadingly. He really *does* want to know things sometimes. Being facetious is not being a very good father. "Sure," I say, reversing myself, "I believe in ghosts too."

Now John and Sylvia look at me peculiarly. I see I'm not going to get out of this one easily and brace myself for a long explanation.

"It's completely natural," I say, "to think of Europeans who believed in ghosts or Indians who believed in ghosts as ignorant. The scientific point of view has wiped out every other view to a point where they all seem primitive, so that if a person today talks about ghosts or spirits he is considered ignorant or maybe nutty. It's just all but completely impossible to imagine a world where ghosts can actually exist."

John nods affirmatively and I continue.

"My own opinion is that the intellect of modern man isn't that superior. IQs aren't that much different. Those Indians and medieval men were just as intelligent as we are, but the context in which they thought was completely different. Within that *context* of thought, ghosts and spirits are quite as real as atoms, particles, photons and quants are to a modern man. In *that* sense I believe in ghosts. Modern man has his ghosts and spirits too, you know."

"What?"

"Oh, the laws of physics and of logic . . . the number system . . . the principle of algebraic substitution. These are ghosts. We just believe in them so thoroughly they seem real."

"They seem real to me," John says.

"I don't get it," says Chris.

So I go on. "For example, it seems completely natural to presume that gravitation and the law of gravitation existed before Isaac Newton. It would sound nutty to think that until the seventeenth century there was no gravity."

"Of course."

"So when did this law start? Has it always existed?"

John is frowning, wondering what I am getting at.

"What I'm driving at," I say, "is the notion that before the beginning of the earth, before the sun and the stars were formed, before the primal generation of anything, the law of gravity existed."

"Sure."

"Sitting there, having no mass of its own, no energy of its own, not in anyone's mind because there wasn't anyone, not in space because there was no space either, not anywhere—this law of gravity still existed?"

Now John seems not so sure.

"If that law of gravity existed," I say, "I honestly don't know what a thing has to do to be *nonexistent*. It seems to me that law of gravity has passed every test of nonexistence there is. You cannot think of a single attribute of nonexistence that that law of gravity didn't have. Or a single scientific attribute of existence it did have. And yet it is still 'common sense' to believe that it existed."

John says, "I guess I'd have to think about it."

"Well, I predict that if you think about it long enough you will find yourself going round and round and round and round until you finally reach only one possible, rational, intelligent conclusion. The law of gravity and gravity itself *did not exist* before Isaac Newton. No other conclusion makes sense.

"And *what that means*," I say before he can interrupt, "and *what that means* is that that law of gravity exists *nowhere* except in people's heads! It's a ghost! We are all of us very arrogant and conceited about running down other people's ghosts but just as ignorant and barbaric and superstitious about our own."

"Why does everybody believe in the law of gravity then?"

"Mass hypnosis. In a very orthodox form known as 'education.' "

"You mean the teacher is hypnotizing the kids into believing the law of gravity?"

"Sure."

"That's absurd."

"You've heard of the importance of eye contact in the classroom? Every educationist emphasizes it. No educationist explains it."

John shakes his head and pours me another drink. He puts his hand over his mouth and in a mock aside says to Sylvia, "You know, most of the time he seems like such a normal guy."

I counter, "That's the first normal thing I've said in weeks. The rest of the time I'm feigning twentieth-century lunacy just like you are. So as not to draw attention to myself.

"But I'll *repeat* it for you," I say. "We believe the disembodied words of Sir Isaac Newton were sitting in the middle of nowhere billions of years before he was born and that magically he *discovered* these words. They were always there, even when they applied to nothing. Gradually the world came into being and then they applied to *it*. In fact, those words themselves were what formed the world. That, John, is ridiculous.

"The problem, the contradiction the scientists are stuck with, is that of *mind*. Mind has no matter or energy but they can't escape its predominance over everything they do. Logic exists in the mind. Numbers exist only in the mind. I don't get upset when scientists say that ghosts exist in the mind. It's that *only* that gets me. Science is *only* in your mind too, it's just that that doesn't make it bad. Or ghosts either."

They are just looking at me so I continue: "Laws of nature are human *inventions,* like ghosts. Laws of logic, of mathematics are also human inventions, like ghosts. The whole blessed thing is a human invention, including the idea that it *isn't* a human invention. The world has no existence whatsoever outside the human imagination. It's all a ghost, and in antiquity was so recognized as a ghost, the whole blessed world we live in. It's run by ghosts. We see what we see because these ghosts *show* it to us, ghosts of Moses and Christ and the Buddha, and Plato, and Descartes, and Rousseau and Jefferson and Lincoln, on and on and on. Isaac Newton is a very good ghost. One of the best. Your common sense is nothing more than the voices of thousands and thousands of these ghosts from the past. Ghosts and more ghosts. Ghosts trying to find their place among the living."

John looks too much in thought to speak. But Sylvia is excited. "Where do you *get* all these ideas?" she asks.

I am about to answer them but then do not. I have a feeling of having already pushed it to the limit, maybe beyond, and it is time to drop it.

After a while John says, "It'll be good to see the mountains again."

"Yes, it will," I agree. "One last drink to that!"

We finish it and are off to our rooms.

I see that Chris brushes his teeth, and let him get by with a promise that he'll shower in the morning. I pull seniority and take the bed by the window. After the lights are out he says, "Now, tell me a ghost story."

"I just did, out there."

"I mean a *real* ghost story."

"That was the realest ghost story you'll ever hear."

"You know what I mean. The other kind."

I try to think of some conventional ones. "I used to know so many of them when I was a kid, Chris, but they're all forgotten," I say. "It's time to go to sleep. We've all got to get up early tomorrow."

Except for the wind through the screens of the motel window it is quiet. The thought of all that wind sweeping toward us across the open fields of the prairie is a tranquil one and I feel lulled by it.

The wind rises and then falls, then rises and sighs, and falls again . . . from so many miles away.

"Did you ever know a ghost?" Chris asks.

I am half asleep. "Chris," I say, "I knew a fellow

once who spent all his whole life doing nothing but hunting for a ghost, and it was just a waste of time. So go to sleep."

I realize my mistake too late.

"Did he find him?"

"Yes, he found him, Chris."

I keep wishing Chris would just listen to the wind and not ask questions.

"What did he do then?"

"He thrashed him good."

"Then what?"

"Then he became a ghost himself." Somehow I had the thought this was going to put Chris to sleep, but it's not and it's just waking me up.

"What is his name?"

"No one you know."

"But what *is* it?"

"It doesn't matter."

"Well, what is it anyway?"

"His name, Chris, since it doesn't matter, is Phaedrus. It's not a name you know."

"Did you see him on the motorcycle in the storm?"

"What makes you say *that*?"

"Sylvia said she thought you saw a ghost."

"That's just an expression."

"Dad?"

"This had better be the last question, Chris, or I'm going to become angry."

"I was just trying to say you sure don't talk like anyone else."

"Yes, Chris, I know that," I say. "It's a problem. Now go to sleep."

"Good night, Dad."

"Good night."

A half hour later he is breathing sleepfully, and the wind is still strong as ever and I am wide-awake. There, out the window in the dark—this cold wind crossing the road into the trees, the leaves shimmering flecks of moonlight—there is no question about it, Phaedrus saw all of this. What he was doing here I have no idea. Why he came this way I will probably never know. But he has been here, steered us onto this strange road, has been with us all along. There is no escape.

I wish I could say that I don't know why he is here, but I'm afraid I must now confess that I do. The ideas, the things I was saying about science and ghosts, and even that idea this afternoon about caring and technology—they are not my own. I haven't really had a new idea in years. They are stolen from him. And he has been watching. And that is why he is here.

With that confession, I hope he will now allow me some sleep.

Poor Chris. "Do you know any ghost stories?" he asked. I could have told him one but even the thought of that is frightening.

I really must go to sleep.

4

Every Chautauqua should have a list somewhere of valuable things to remember that can be kept in some safe place for times of future need and inspiration. Details. And now, while the others are still snoring away wasting this beautiful morning sunlight . . . well . . . to sort of fill time . . .

What I have here is my list of valuable things to take on your next motorcycle trip across the Dakotas.

I've been awake since dawn. Chris is still sound asleep in the other bed. I started to roll over for more sleep but heard a rooster crowing and then became aware we are on vacation and there is no point in sleeping. I can hear John right through the motel partition sawing wood in there . . . unless it's Sylvia . . . no, that's too loud. Damned *chain* saw, it sounds like. . . .

I got so tired of forgetting things on trips like this, I made this up and store it in a file at home to check off when I am ready to go.

Most of the items are commonplace and need no

comment. Some of them are peculiar to motorcycling and need some comment. Some of them are just plain peculiar and need a lot of comment. The list is divided into four parts: Clothing, Personal Stuff, Cooking and Camping Gear, and Motorcycle Stuff.

The first part, Clothing, is simple:

1. Two changes of underwear.
2. Long underwear.
3. One change of shirt and pants for each of us. I use Army-surplus fatigues. They're cheap, tough and don't show dirt. I had an item called "dress clothes" at first but John penciled "Tux" after this item. I was just thinking of something you might want to wear outside a filling station.
4. One sweater and jacket each.
5. Gloves. Unlined leather gloves are best because they prevent sunburn, absorb sweat and keep your hands cool. When you're going for an hour or two little things like this aren't important, but when you're going all day long day after day they become plenty important.
6. Cycle boots.
7. Rain gear.
8. Helmet and sunshade.
9. Bubble. This gives me claustrophobia, so I use it only in the rain, which otherwise at high speed stings your face like needles.
10. Goggles. I don't like windshields because they also close you in. These are some British laminated plateglass goggles that

work fine. The wind gets behind sunglasses.
Plastic goggles get scratched up and distort
vision.

The next list is Personal Stuff:
Combs. Billfold. Pocketknife. Memoranda booklet.
Pen. Cigarettes and matches. Flashlight. Soap and
plastic soap container. Toothbrushes and toothpaste.
Scissors. APCs for headaches. Insect repellent. De-
odorant (after a hot day on a cycle, your best friends
don't need to tell you). Sunburn lotion. (On a cycle you
don't notice sunburn until you stop, and then it's too
late. Put it on early.) Band-Aids. Toilet paper. Wash-
cloth (this can go into a plastic box to keep other stuff
from getting damp). Towel.

Books. I don't know of any other cyclist who takes
books with him. They take a lot of space, but I have
three of them here anyway, with some loose sheets of
paper in them for writing. These are:

a. The shop manual for this cycle.
b. A general troubleshooting guide containing
   all the technical information I can never keep
   in my head. This is *Chilton's Motorcycle
   Troubleshooting Guide* written by Ocee Rich
   and sold by Sears, Roebuck.
c. A copy of Thoreau's *Walden* . . . which Chris
   has never heard and which can be read a hundred times without exhaustion. I try always to
   pick a book far over his head and read it as a
   basis for questions and answers, rather than
   without interruption. I read a sentence or two,
   wait for him to come up with his usual barrage

of questions, answer them, then read another sentence or two. Classics read well this way. They must be written this way. Sometimes we have spent a whole evening reading and talking and discovered we have only covered two or three pages. It's a form of reading done a century ago . . . when Chautauquas were popular. Unless you've tried it you can't imagine how pleasant it is to do it this way.

I see Chris is sleeping over there completely relaxed, none of his normal tension. I guess I won't wake him up yet.

Camping Equipment includes:

1. Two sleeping bags.
2. Two ponchos and one ground cloth. These convert into a tent and also protect the luggage from rain while you are traveling.
3. Rope.
4. U. S. Geodetic Survey maps of an area where we hope to do some hiking.
5. Machete.
6. Compass.
7. Canteen. I couldn't find this anywhere when we left. I think the kids must have lost it somewhere.
8. Two Army-surplus mess kits with knife, fork and spoon.
9. A collapsible Sterno stove with one medium-sized can of Sterno. This is an experimental purchase. I haven't used it yet. When it rains

or when you're above the timberline fire-
wood is a problem.
10. Some aluminum screw-top tins. For lard,
salt, butter, flour, sugar. A mountaineering
supply house sold us these years ago.
11. Brillo, for cleaning.
12. Two aluminum-frame backpacks.

Motorcycle Stuff. A standard tool kit comes with the
cycle and is stored under the seat. This is supplemented with the following:

A large, adjustable open-end wrench. A machinist's hammer. A cold chisel. A taper punch. A
pair of tire irons. A tire-patching kit. A bicycle
pump. A can of molybdenum disulfide spray for
the chain. (This has tremendous penetrating ability into the inside of each roller where it really
counts, and the lubricating superiority of molybdenum disulfide is well known. Once it has dried
off, however, it ought to be supplemented with
good old SAE-30 engine oil.) Impact driver. A
point file. Feeler gauge. Test lamp.

Spare parts include:

Plugs. Throttle, clutch and brake cables.
Points, fuses, headlight and taillight bulbs, chain-
coupling link with keeper, cotter pins, baling
wire. Spare chain (this is just an old one that was
about shot when I replaced it, enough to get to a
cycle shop if the present one goes).

And that's about it. No shoelaces.

It would probably be normal about this time to wonder what sort of U-Haul trailer all this is in. But it's not as bulky, really, as it sounds.

I'm afraid these other characters will sleep all day if I let them. The sky outside is sparkling and clear, it's a shame to waste it like this.

I go over finally and give Chris a shake. His eyes pop open, then he sits bolt upright uncomprehending.

"Shower time," I say.

I go outside. The air is invigorating. In fact— Christ!—it is *cold* out. I pound on the Sutherlands' door.

"Yahp," comes John's sleepy voice through the door. "Um-hmmmm. Yahp."

It feels like *autumn*. The cycles are wet with dew. No rain today. But *cold*! It must be in the forties.

While waiting I check the engine oil level and tires, and bolts, and chain tension. A little slack there, and I get out the tool kit and tighten it up. I'm really getting anxious to get going.

I see that Chris dresses warmly and we are packed and on the road, and it is definitely cold. Within minutes all the heat of the warm clothing is drained out by the wind and I am shivering with big shivers. Bracing.

It ought to warm up as soon as the sun gets higher in the sky. About half an hour of this and we'll be in Ellendale for breakfast. We should cover a lot of miles today on these straight roads.

If it weren't so damn cold this would be just gorgeous riding. Low-angled dawn sun striking what looks almost like frost covering those fields, but I guess it's just dew, sparkling and kind of misty. Dawn shadows everywhere make it look less flat than yesterday. All to ourselves. Nobody's even up yet, it looks like. My watch says six-thirty. The old glove above it looks like it's got frost on it, but I guess it's just residues from the soaking last night. Good old beat-up gloves. They are so stiff now from the cold I can hardly straighten my hand out.

I talked yesterday about caring, I *care* about these moldy old riding gloves. I smile at them flying through the breeze beside me because they have been there for so many years and are so old and so tired and so rotten there is something kind of humorous about them. They have become filled with oil and sweat and dirt and spattered bugs and now when I set them down flat on a table, even when they are not cold, they won't stay flat. They've got a memory of their own. They cost only three dollars and have been restitched so many times it is getting impossible to repair them, yet I take a lot of time and pains to do it anyway because I can't imagine any new pair taking their place. That is impractical, but practicality isn't the whole thing with gloves or with anything else.

The machine itself receives some of the same feelings. With over 27,000 on it it's getting to be something of a high-miler, an old-timer, although there are plenty of older ones running. But over the miles, and I think most cyclists will agree with this, you pick up certain feelings about an individual machine that are unique for that one individual machine and no other. A friend who owns a cycle of the same make, model and even same year brought it over for repair, and when I test rode it afterward it was hard to believe it had come

from the same factory years ago. You could see that
long ago it had settled into its own kind of feel and ride
and sound, completely different from mine. No worse,
but different.

I suppose you could call that a personality. Each machine *has* its own, unique personality which probably
could be defined as the intuitive sum total of everything you know and feel about it. This personality constantly changes, usually for the worse, but sometimes
surprisingly for the better, and it is this personality that
is the real object of motorcycle maintenance. The new
ones start out as good-looking strangers and, depending on how they are treated, degenerate rapidly into
bad-acting grouches or even cripples, or else turn into
healthy, good-natured, long-lasting friends. This one,
despite the murderous treatment it got at the hands of
those alleged mechanics, seems to have recovered and
has been requiring fewer and fewer repairs as time
goes on.

There it is! Ellendale!

A water tower, groves of trees and buildings among
them in the morning sunlight. I've just given in to the
shivering which has been almost continuous the whole
trip. The watch says seven-fifteen.

A few minutes later we park by some old brick
buildings. I turn to John and Sylvia who have pulled up
behind us. "That was *cold*!" I say.

They just stare at me fish-eyed.

"Bracing, what?" I say. No answer.

I wait until they are completely off, then see that
John is trying to untie all their luggage. He is having
trouble with the knot. He gives up and we all move
toward the restaurant.

I try again, I'm walking backward in front of them toward the restaurant, feeling a little manic from the ride, wringing my hands and laughing. "Sylvia! Speak to me!" Not a smile.

I guess they really *were* cold.

They order breakfast without looking up.

Breakfast ends, and I say finally, "What next?"

John says slowly and deliberately. "We're not leaving here until it warms up." He has a sheriff-at-sundown tone in his voice, which I suppose makes it final.

So John and Sylvia and Chris sit and stay warm in the lobby of the hotel adjoining the restaurant, while I go out for a walk.

I guess they're kind of mad at me for getting them up so early to ride through that kind of stuff. When you're stuck together like this, I figure small differences in temperament are bound to show up. I remember, now that I think of it, I've never been cycling with them before one or two o'clock in the afternoon, although for me dawn and early morning is always the greatest time for riding.

The town is clean and fresh and unlike the one we woke up in this morning. Some people are on the street and are opening stores and saying, "Good morning" and talking and commenting about how cold it is. Two thermometers on the shady side of the street read 42 and 46 degrees. One in the sun reads 65 degrees.

After a few blocks the main street goes onto two hard, muddy tracks into a field, past a quonset hut full of farm machinery and repair tools, and then ends in a field. A man standing in the field is looking at me suspiciously, wondering what I am doing, probably, as I look into the quonset hut. I return down the street, find a chilly bench and stare at the motorcycle. Nothing to do.

It was cold all right, but not *that* cold. How do John and Sylvia ever get through Minnesota winters? I wonder. There's kind of a glaring inconsistency here, that's almost too obvious to dwell on. If they can't stand physical discomfort and they can't stand technology, they've got a little compromising to do. They depend on technology and condemn it at the same time. I'm sure they know that and that just contributes to their dislike of the whole situation. They're not presenting a logical thesis, they're just reporting how it is. But three farmers are coming into town now, rounding the corner in that brand-new pickup truck. I'll bet with them it's just the other way around. They're going to show off that truck and their tractor and that new washing machine and they'll have the tools to fix them if they go wrong, and know how to use the tools. They *value* technology. And they're the ones who need it the *least*. If all technology stopped, tomorrow, these people would know how to make out. It would be rough, but they'd survive. John and Sylvia and Chris and I would be dead in a week. This condemnation of technology is ingratitude, that's what it is.

Blind alley, though. If someone's ungrateful and you tell him he's ungrateful, okay, you've called him a name. You haven't solved anything.

A half hour later the thermometer by the hotel door reads 53 degrees. Inside the empty main dining room of the hotel I find them, looking restless. They seem, by their expressions, to be in a better mood though, and John says optimistically, "I'm going to put on everything I own, and then we'll make it all right."

He goes out to the cycles, and when he comes back says, "I sure hate to unpack all that stuff, but I don't want another ride like that last one." He says it is freezing in the men's room, and since there is no one else in the dining room, he crosses behind a table back from where we are sitting, and I am sitting at the table, talking to Sylvia, and then I look over and there is John, all decked out in a full-length set of pale-blue long underwear. He is smirking from ear to ear at how silly he looks. I stare at his glasses lying on the table for a moment and then say to Sylvia:

"You know, just a moment ago we were sitting here talking to Clark Kent . . . see, there's his glasses . . . and now all of a sudden . . . Lois, do you suppose? . . ."

John howls. "CHICKENMAN!"

He glides over the varnished lobby floor like a skater, does a handspring, then glides back. He raises one arm over his head and then crouches as if starting for the sky. "I'm ready, here I go!" He shakes his head sadly. "Jeez, I hate to bust through that nice ceiling, but my X-ray vision tells me somebody's in trouble." Chris is giggling.

"We'll all be in trouble if you don't get some clothes on," Sylvia says.

John laughs. "An exposer, hey? 'The Ellendale revealer!' " He struts around some more, then begins to put his clothes on over the underwear. He says, "Oh no, oh no, they wouldn't do that. Chickenman and the police have an understanding. They know who's on the side of law and order and justice and decency and fair play for everyone."

When we hit the highway again it is still chilly, but not like it was. We pass through a number of towns and gradually, almost imperceptibly, the sun warms us up, and my feelings warm up with it. The tired feeling wears off completely and the wind and sun feel good now, making it real. It's happening, just from the warming of the sun, the road and green prairie farmland and buffeting wind coming together. And soon it is nothing but beautiful warmth and wind and speed and sun down the empty road. The last chills of the morning are thawed by the warm air. Wind and more sun and more smooth road.

So green this summer and so fresh.

There are white and gold daisies among the grass in front of an old wire fence, a meadow with some cows and far in the distance a low rising of the land with something golden on it. Hard to know what it is. No need to know.

Where there is a slight rise in the road the drone of the motor becomes heavier. We top the rise, see a new spread of land before us, the road descends and the drone of the engine falls away again. Prairie. Tranquil and detached.

Later, when we stop, Sylvia has tears in her eyes from the wind, and she stretches out her arms and says, "It's so beautiful. It's so empty."

I show Chris how to spread his jacket on the ground and use an extra shirt for a pillow. He is not at all sleepy but I tell him to lie down anyway, he'll need the rest. I open up my own jacket to soak up more heat. John gets his camera out.

After a while he says, "This is the hardest stuff in the world to photograph. You need a three-hundred-and-sixty-degree lens, or something. You see it, and then you look down in the ground glass and it's just nothing. As soon as you put a border on it, it's gone."

I say, "That's what you don't see in a car, I suppose."

Sylvia says, "Once when I was about ten we stopped like this by the road and I used half a roll of film taking pictures. And when the pictures came back I cried. There wasn't anything there."

"When are we going to get going?" Chris says.

"What's your hurry?" I ask.

"I just want to get going."

"There's nothing up ahead that's any better than it is right here."

He looks down silently with a frown. "Are we going to go camping tonight?" he asks. The Sutherlands look at me apprehensively.

"Are we?" he repeats.

"We'll see later," I say.

"Why later?"

"Because I don't know now."

"Why don't you know now?"

"Well, I just don't know now why I just don't know."

John shrugs that it's okay.

"This isn't the best camping country," I say. "There's no cover and no water." But suddenly I add, "All right, tonight we'll camp out." We had talked about it before.

So we move down the empty road. I don't want to own these prairies, or photograph them, or change them, or even stop or even keep going. We are just moving down the empty road.

5

The flatness of the prairie disappears and a deep undulation of the earth begins. Fences are rarer, and the greenness has become paler . . . all signs that we approach the High Plains.

We stop for gas at Hague and ask if there is any way to get across the Missouri between Bismarck and Mobridge. The attendant doesn't know of any. It is hot now, and John and Sylvia go somewhere to get their long underwear off. The motorcycle gets a change of oil and chain lubrication. Chris watches everything I do but with some impatience. Not a good sign.

"My eyes hurt," he says.

"From what?"

"From the wind."

"We'll look for some goggles."

All of us go in a shop for coffee and rolls. Everything is different except one another, so we look around rather than talk, catching fragments of conversation among people who seem to know each other and are glancing at *us* because *we're* new. Afterward, down the street, I find a thermometer for storage in the saddlebags and some plastic goggles for Chris.

The hardware man doesn't know any short route across the Missouri either. John and I study the map. I had hoped we might find an unofficial ferryboat crossing or footbridge or something in the ninety-mile stretch, but evidently there isn't any because there's not much to get to on the other side. It's all Indian reservation. We decide to head south to Mobridge and cross there.

The road south is awful. Choppy, narrow, bumpy concrete with a bad head wind, going into the sun and big semis going the other way. These roller-coaster hills speed them up on the down side and slow them up on the up side and prevent our seeing very far ahead, making passing nervewracking. The first one gave me a scare because I wasn't ready for it. Now I hold tight and brace for them. No danger. Just a shock wave that hits you. It is hotter and dryer.

At Herreid John disappears for a drink while Sylvia and Chris and I find some shade in a park and try to rest. It isn't restful. A change has taken place and I don't know quite what it is. The streets of this town are broad, much broader than they need be, and there is a pallor of dust in the air. Empty lots here and there between the buildings have weeds growing in them. The sheet metal equipment sheds and water tower are like those of previous towns but more spread out. Everything is more run-down and mechanical-looking, and sort of randomly located. Gradually I see what it is. Nobody is concerned anymore about tidily conserving space. The land isn't valuable anymore. We are in a Western town.

We have lunch of hamburgers and malts at an A & W place in Mobridge, cruise down a heavily trafficked main street and then there it is, at the bottom of the hill, the Missouri. All that moving water is strange, banked by grass hills that hardly get any water at all. I turn around and glance at Chris but he doesn't seem to be particularly interested in it.

We coast down the hill, clunk onto the bridge and across we go, watching the river through the girders moving by rhythmically, and then we are on the other side.

We climb a long, long hill into another kind of country.

The fences are really all gone now. No brush, no trees. The sweep of the hills is so great John's motorcycle looks like an ant up ahead moving through the green slopes. Above the slopes outcroppings of rocks stand out overhead at the tops of the bluffs.

It all has a natural tidiness. If it were abandoned land there would be a chewed-up, scruffy look, with chunks of old foundation concrete, scraps of painted sheet metal and wire, weeds that had gotten in where the sod was broken up for whatever little enterprise was attempted. None of that here. Not kept up, just never messed up in the first place. It's just the way it always must have been. Reservation land.

There's no friendly motorcycle mechanic on the other side of those rocks and I'm wondering if we're ready for this. If anything goes wrong now we're in real trouble.

I check the engine temperature with my hand. It's reassuringly cool. I put in the clutch and let it coast for a second in order to hear it idling. Something sounds funny and I do it again. It takes a while to figure out that it's not the engine at all. There's an echo from the bluff ahead that lingers after the throttle is closed. Funny. I do this two or three times. Chris wonders what's wrong and I have him listen to the echo. No comment from him.

This old engine has a nickels-and-dimes sound to it. As if there were a lot of loose change flying around inside. Sounds awful, but it's just normal valve clatter. Once you get used to that sound and learn to expect it, you automatically hear any difference. If you don't hear any, that's good.

I tried to get John interested in that sound once but it was hopeless. All he heard was noise and all he saw was the machine and me with greasy tools in my hands, nothing else. That didn't work.

He didn't really see what was going on and was not interested enough to find out. He isn't so interested in what things *mean* as in what they *are*. That's quite important, that he sees things this way. It took me a long time to see this difference and it's important for the Chautauqua that I make this difference clear.

I was so baffled by his refusal even to think about any mechanical subject I kept searching for ways to clue him to the whole thing but didn't know where to start.

I thought I would wait until something went wrong with *his* machine and then I would help *him* fix it and *that* way get him into it, but I goofed that one myself because I didn't understand this difference in the way he looked at things.

His handlebars had started flipping. Not badly, he said, just a little when you shoved hard on them. I warned him not to use his adjustable wrench on the tightening nuts. It was likely to damage the chrome and start small rust spots. He agreed to use my metric sockets and box-ends.

When he brought his motorcycle over I got my wrenches out but then noticed that no amount of tightening would stop the slippage, because the ends of the collars were pinched shut.

"You're going to have to shim those out," I said.

"What's shim?"

"It's a thin, flat strip of metal. You just slip it around the handlebar under the collar there and it will open up the collar to where you can tighten it again. You use shims like that to make adjustments in all kinds of machines."

"Oh," he said. He was getting interested. "Good. Where do you buy them?"

"I've got some right here," I said gleefully, holding up a can of beer in my hand.

He didn't understand for a moment. Then he said, "*What*, the *can*?"

"Sure," I said, "best shim stock in the world."

I thought this was pretty clever myself. Save him a trip to God knows where to get shim stock. Save him time. Save him money.

But to my surprise he didn't see the cleverness of this at all. In fact he got noticeably haughty about the whole thing. Pretty soon he was dodging and filling with all kinds of excuses and, before I realized what his real attitude was, we had decided not to fix the handlebars after all.

As far as I know those handlebars are still loose. And I believe now that he was actually offended at the time. I had had the *nerve* to propose repair of his new eighteen-hundred-dollar BMW, the pride of a half-century of German mechanical finesse, with a piece of old *beer* can!

Ach, du lieber!

Since then we have had very few conversations about motorcycle maintenance. None, now that I think of it.

You push it any further and suddenly you are angry, without knowing why.

I should say, to explain this, that beer-can aluminum is soft and sticky, as metals go. Perfect for the application. Aluminum doesn't oxidize in wet weather—or, more precisely, it always has a thin layer of oxide that prevents any further oxidation. Also perfect.

In other words, any *true* German mechanic, with a half-century of mechanical finesse behind him, would have concluded that this particular solution to this particular technical problem was *perfect*.

For a while I thought what I *should* have done was sneak over to the workbench, cut a shim from the beer can, remove the printing and then come back and tell him we were in luck, it was the last one I had, specially imported from Germany. That would have done it. A special shim from the private stock of Baron Alfred Krupp, who had to sell it at a great sacrifice. Then he would have gone gaga over it.

That Krupp's-private-shim fantasy gratified me for a while, but then it wore off and I saw it was just being vindictive. In its place grew that old feeling I've talked about before, a feeling that there's something bigger involved than is apparent on the surface. You follow these little discrepancies long enough and they sometimes open up into huge revelations. There was just a feeling on my part that this was something a little bigger than I wanted to take on without thinking about it, and I turned instead to my usual habit of trying to extract causes and effects to see what was involved that could possibly lead to such an impasse between John's view of that lovely shim and my own. This comes up all the time in mechanical work. A hang-up. You just sit and stare and think, and search randomly for new information, and go away and come back again, and after a while the unseen factors start to emerge.

What emerged in vague form at first and then in sharper outline was the explanation that I had been seeing that shim in a kind of intellectual, rational, cerebral way in which the scientific properties of the metal were all that counted. John was going at it immediately and intuitively, grooving on it. I was going at it in terms of underlying form. He was going at it in terms of immediate appearance. I was seeing what the shim *meant*. He was seeing what the shim *was*. That's how I arrived at that distinction. And when you see what the shim *is,* in this case, it's depressing. Who likes to think of a beautiful precision machine fixed with an old hunk of junk?

I guess I forgot to mention John is a musician, a drummer, who works with groups all over town and makes a pretty fair income from it. I suppose he just thinks about everything the way he thinks about drumming—which is to say he doesn't really *think* about it at all. He just does it. Is with it. He just responded to fixing his motorcycle with a beer can the way he would respond to someone dragging the beat while he was playing. It just did a big thud with him and that was it. He didn't want any part of it.

At first this difference seemed fairly minor, but then it grew . . . and grew . . . and grew . . . until I began to see why I missed it. Some things you miss because they're so tiny you overlook them. But some things you don't see because they're so *huge.* We were both looking at the same thing, seeing the same thing, talking about the same thing, thinking about the same thing, except he was looking, seeing, talking and thinking from a completely different *dimension.*

He really *does* care about technology. It's just that in this other dimension he gets all screwed up and is rebuffed by it. It just won't swing for him. He tries to swing it without any rational premeditation and botches it and botches it and botches it and after so many botches gives up and just kind of puts a blanket curse on that whole nuts-and-bolts scene. He will not or cannot believe there is anything in this world for which grooving is not the way to go.

That's the dimension he's in. The groovy dimension. I'm being awfully square talking about all this mechanical stuff all the time. It's all just parts and relationships and analyses and syntheses and figuring things out and it isn't really *here.* It's somewhere else, which thinks it's here, but's a million miles away. *This* is what it's all about. He's on this dimensional difference which underlay much of the cultural changes of the sixties, I think, and is still in the process of reshaping our whole national outlook on things. The "generation gap" has been a result of it. The names "beat" and "hip" grew out of it. Now it's become apparent that this dimension isn't a fad that's going to go away next year or the year after. It's here to stay because it's a very serious and important way of looking at things that *looks* incompatible with reason and order and responsibility but actually is not. Now we are down to the root of things.

My legs have become so stiff they are aching. I hold them out one at a time and turn my foot as far to the left and to the right as it will go to stretch the leg. It helps, but then the other muscles get tired from holding the legs out.

What we have here is a conflict of *visions of reality*. The world as you see it right here, right now, *is reality,* regardless of what the scientists say it might be. That's the way John sees it. But the world as revealed by its scientific discoveries is also reality, regardless of how it may appear, and people in John's dimension are going to have to do more than just ignore it if they want to hang on to their vision of reality. John will discover this if his points burn out.

That's really why he got upset that day when he couldn't get his engine started. It was an *intrusion on his reality*. It just blew a hole right through his whole groovy way of looking at things and he would not face up to it because it seemed to threaten his whole life style. In a way he was experiencing the same sort of anger scientific people have sometimes about abstract art, or at least *used* to have. That didn't fit *their* life style either.

What you've got here, really, are *two* realities, one of immediate artistic appearance and one of underlying scientific explanation, and they don't match and they don't fit and they don't really have much of anything to do with one another. That's quite a situation. You might say there's a little problem here.

At one stretch in the long desolate road we see an isolated grocery store. Inside, in back, we find a place to sit on some packing cases and drink canned beer.

The fatigue and backache are getting to me now. I push the packing case over to a post and lean on that.

Chris's expression shows he is really settling into something bad. This has been a long hard day. I told Sylvia way back in Minnesota that we could expect a slump in spirits like this on the second or third day and now it's here. Minnesota—when was that?

A woman, badly drunk, is buying beer for some man she's got outside in a car. She can't make up her mind what brand to buy and the wife of the owner waiting on her is getting mad. She still can't decide, but then sees us, and weaves over and asks if we own the motorcycles. We nod yes. Then she wants a ride on one. I move back and let John handle this.

He puts her off graciously, but she comes back again and again, offering him a dollar for a ride. I make some jokes about it, but they're not funny and just add to the depression. We get out and back into the brown hills and heat again.

By the time we reach Lemmon we are really aching tired. At a bar we hear about a campground to the south. John wants to camp in a park in the middle of Lemmon, a comment that sounds strange and angers Chris greatly.

I'm more tired now than I can remember having been in a long time. The others too. But we drag ourselves through a supermarket, pick up whatever groceries come to mind and with some difficulty pack them onto the cycles. The sun is so far down we're running out of light. It'll be dark in an hour. We can't seem to get moving. I wonder, are we dawdling, or what?

"C'mon, Chris, let's go," I say.

"Don't holler at *me*. *I'm* ready."

We drive down a county road from Lemmon, exhausted, for what seems a long, long time, but can't be too long because the sun is still above the horizon. The campsite is deserted. Good. But there is less than a half-hour of sun and no energy left. This is the hardest now.

I try to get unpacked as fast as possible but am so stupid with exhaustion I just set everything by the camp road without seeing what a bad spot it is. Then I see it is too windy. This is a High Plains wind. It is semidesert here, everything burned up and dry except for a lake, a large reservoir of some sort below us. The wind blows from the horizon across the lake and hits us with sharp gusts. It is already chilly. There are some scrubby pines back from the road about twenty yards and I ask Chris to move the stuff over there.

He doesn't do it. He wanders off down to the reservoir. I carry the gear over by myself.

I see between trips that Sylvia is making a real effort at setting things up for cooking, but she's as tired as I am.

The sun goes down.

John has gathered wood but it's too big and the wind is so gusty it's hard to start. It needs to be splintered into kindling. I go back over the scrub pines, hunt around through the twilight for the machete, but it's already so dark in the pines I can't find it. I need the flashlight. I look for it, but it's too dark to find that either.

I go back and start up the cycle and ride it back over to shine the headlight on the stuff so that I can find the flashlight. I look through all the stuff item by item to find the flashlight. It takes a long time to realize I don't need the flashlight, I need the machete, which is in plain sight. By the time I get it back John has got the fire going. I use the machete to hack up some of the larger pieces of wood.

Chris reappears. *He's* got the flashlight!

"When are we going to eat?" he complains.

"We're getting it fixed as fast as possible," I tell him. "Leave the flashlight here."

He disappears again, taking the flashlight with him.

The wind blows the fire so hard it doesn't reach up to cook the steaks. We try to fix up a shelter from the wind using large stones from the road, but it's too dark to see what we're doing. We bring both cycles over and catch the scene in a crossbeam of headlights. Peculiar light. Bits of ash blowing up from the fire suddenly glow bright white in it, then disappear in the wind.

BANG! There's a loud explosion behind us. Then I hear Chris giggling.

Sylvia is upset.

"I found some firecrackers," Chris says.

I catch my anger in time and say to him, coldly, "It's time to eat now."

"I need some matches," he says.

"Sit down and eat."

"Give me some matches first."

"Sit down and eat."

He sits down and I try to eat the steak with my Army mess knife, but it is too tough, and so I get out a hunting knife and use it instead. The light from the motorcycle headlight is full upon me so that the knife, when it goes down into the mess gear, is in full shadow and I can't see where it's going.

Chris says he can't cut his either and I pass my knife to him. While reaching for it he dumps everything onto the tarp.

No one says a word.

I'm not angry that he spilled it, I'm angry that now the tarp's going to be greasy the rest of the trip.

"Is there any more?" he asks.

"Eat *that*," I say. "It just fell on the tarp."

"It's too dirty," he says.

"Well, that's all there is."

A wave of depression hits. I just want to go to sleep now. But he's angry and I expect we're going to have one of his little scenes. I wait for it and pretty soon it starts.

"I don't like the taste of this," he says.

"Yes, that's rough, Chris."

"I don't like *any* of this. I don't like this camping at all."

"It was *your* idea," Sylvia says. "You're the one who wanted to go camping."

She shouldn't say that, but there's no way she can know. You take his bait and he'll feed you another one, and then another, and another until you finally hit him, which is what he really wants.

"I don't care," he says.

"Well, you ought to," she says.

"Well, I don't."

An explosion point is very near. Sylvia and John look at me but I remain deadpan. I'm sorry about this but there's nothing I can do right now. Any argument will just worsen things.

"I'm not hungry," Chris says.

No one answers.

"My stomach hurts," he says.

The explosion is avoided when Chris turns and walks away in the darkness.

We finish eating. I help Sylvia clean up, and then we sit around for a while. We turn the cycle lights off to conserve the batteries and because the light from them is ugly anyway. The wind has died down some and there is a little light from the fire. After a while my eyes become accustomed to it. The food and anger have taken off some of the sleepiness. Chris doesn't return.

"Do you suppose he's just *punishing*?" Sylvia asks.

"I suppose," I say, "although it doesn't sound quite right." I think about it and add, "That's a child-psychology term—a context I dislike. Let's just say he's being a complete bastard."

John laughs a little.

"Anyway," I say, "it was a good supper. I'm sorry he had to act up like this."

"Oh, that's all right," John says. "I'm just sorry he won't get anything to eat."

"It won't hurt him."

"You don't suppose he'll get lost out there."

"No, he'll holler if he is."

Now that he has gone and we have nothing to do I become more aware of the space all around us. There is not a sound anywhere. Lone prairie.

Sylvia says, "Do you suppose he really has stomach pains?"

"Yes," I say, somewhat dogmatically. I'm sorry to see the subject continued but they deserve a better explanation than they're getting. They probably sense that there's more to it than they've heard. "I'm sure he does," I finally say. "He's been examined a half-dozen times for it. Once it was so bad we thought it was appendicitis. . . . I remember we were on a vacation up north. I'd just finished getting out an engineering proposal for a five-million-dollar contract that just about did me in. That's a whole other world. No time and no patience and six hundred pages of information to get out the door in one week and I was about ready to kill three different people and we thought we'd better head for the woods for a while.

"I can hardly remember what part of the woods we were in. Head just spinning with engineering data, and anyway Chris was just screaming. We couldn't touch him, until I finally saw I was going to have to pick him up fast and get him to the hospital, and where that was I'll never remember, but they found nothing."

"Nothing?"

"No. But it happened again on other occasions too."

"Don't they have *any* idea?" Sylvia asks.

"This spring they diagnosed it as the beginning symptoms of mental illness."

"What?" John says.

It's too dark to see Sylvia or John now or even the outlines of the hills. I listen for sounds in the distance, but hear none. I don't know what to answer and so say nothing.

When I look hard I can make out stars overhead but

the fire in front of us makes it hard to see them. The night all around is thick and obscure. My cigarette is down to my fingers and I put it out.

"I didn't know that," Sylvia's voice says. All traces of anger are gone. "We wondered why you brought him instead of your wife," she says. "I'm glad you told us."

John shoves some of the unburned ends of the wood into the fire.

Sylvia says, "What do you suppose the cause is?"

John's voice rasps, as if to cut it off, but I answer, "I don't know. Causes and effects don't seem to fit. Causes and effects are a result of thought. I would think mental illness comes before thought." This doesn't make sense to them, I'm sure. It doesn't make much sense to me and I'm too tired to try to think it out and give it up.

"What do the psychiatrists think?" John asks.

"Nothing. I stopped it."

"Stopped it?"

"Yes."

"Is that good?"

"I don't know. There's no rational reason I can think of for saying it's *not* good. Just a mental block of my own. I think about it and all the good reasons for it and make plans for an appointment and even look for the phone number and then the block hits, and it's just like a door slammed shut."

"That doesn't sound right."

"No one else thinks so either. I suppose I can't hold out forever."

"But *why*?" Sylvia asks.

"I don't *know* why . . . it's just that . . . I don't

know . . . they're not *kin*." . . . Surprising word, I think to myself never used it before. Not of *kin* . . . sounds like hillbilly talk . . . not of a *kind* . . . same root . . . *kind*ness, too . . . they can't have real *kind*ness toward him, they're not his *kin*. . . . That's exactly the feeling.

Old word, so ancient it's almost drowned out. What a change through the centuries. Now anybody can be "kind." And everybody's supposed to be. Except that long ago it was something you were born into and couldn't help. Now it's just a faked-up attitude half the time, like teachers the first day of class. But what do they really know about *kind*ness who are not *kin*?

It goes over and over again through my thoughts . . . *mein Kind*—my child. There it is in another language. *Mein Kinder* . . . *"Wer reitet so spät durch Nacht und Wind? Es ist der Vater mit seinem Kind."*

Strange feeling from that.

"What are you thinking about?" Sylvia asks.

"An old poem, by Goethe. It must be two hundred years old. I had to learn it a long time ago. I don't know why I should remember it now, except . . ." The strange feeling comes back.

"How does it go?" Sylvia asks.

I try to recall. "A man is riding along a beach at night, through the wind. It's a father, with his son, whom he holds fast in his arm. He asks his son why he looks so pale, and the son replies, 'Father, don't you see the ghost?' The father tried to reassure the boy it's only a bank of fog along the beach that he sees and only the rustling of the leaves in the wind that he hears but the son keeps saying it is the ghost and the father rides harder and harder through the night."

"How does it end?"

"In failure . . . death of the child. The ghost wins."

The wind blows light up from the coals and I see Sylvia look at me startled.

"But that's another land and another time," I say. "Here life is the end and ghosts have no meaning. I believe that. I believe in all this too," I say, looking out at the darkened prairie, "although I'm not sure of what it all means yet . . . I'm not sure of much of anything these days. Maybe that's why I talk so much."

The coals die lower and lower. We smoke our cigarettes. Chris is off somewhere in the darkness but I'm not going to shag after him. John is carefully silent and Sylvia is silent and suddenly we are all separate, all alone in our private universes, and there is no communication among us. We douse the fire and go back to the sleeping bags in the pines.

I discover that this one tiny refuge of scrub pines where I have put the sleeping bags is also the refuge from the wind of millions of mosquitoes up from the reservoir. The mosquito repellent doesn't stop them at all. I crawl deep into the sleeping bag and make one little hole for breathing. I am almost asleep when Chris finally shows up.

"There's a great big sandpile over there," he says, crunching around on the pine needles.

"Yes," I say. "Get to sleep."

"You should see it. Will you come and see it tomorrow?"

"We won't have time."

"Can I play over there tomorrow morning?"

"Yes."

He makes interminable noises getting undressed and into the sleeping bag. He is in it. Then he rolls around.

Then he is silent, and then rolls some more. Then he says, "Dad?"

"What?"

"What was it like when you were a kid?"

"Go to *sleep,* Chris!" There are limits to what you can listen to.

Later I hear a sharp inhaling of phlegm that tells me he has been crying, and though I'm exhausted, I don't sleep. A few words of consolation might have helped there. He was trying to be friendly. But the words aren't forthcoming for some reason. Consoling words are more for strangers, for hospitals, not kin. Little emotional Band-Aids like that aren't what he needs or what's sought. . . . I don't know what he needs, or what's sought.

A gibbous moon comes up from the horizon beyond the pines, and by its slow, patient arc across the sky I measure hour after hour of semisleep. Too much fatigue. The moon and strange dreams and sounds of mosquitoes and odd fragments of memory become jumbled and mixed in an unreal lost landscape in which the moon is shining and yet there is a bank of fog and I am riding a horse and Chris is with me and the horse jumps over a small stream that runs through the sand toward the ocean somewhere beyond. And then that is broken. . . . And then it reappears.

And in the fog there appears an intimation of a figure. It disappears when I look at it directly, but then reappears in the corner of my vision when I turn my glance. I am about to say something, to call to it, to recognize it, but then do not, knowing that to recognize it by any gesture or action is to give it a reality which it must not have. But it is a figure I recognize even though I do not let on. It is Phaedrus.

Evil spirit. Insane. From a world without life or death.

The figure fades and I hold panic down . . . tight . . . not rushing it . . . just letting it sink in . . . not believing it, not disbelieving it . . . but the hair crawls slowly on the back of my skull . . . he is calling Chris, is that it? . . . Yes? . . .

6

My watch says nine o'clock. And it's already too hot to sleep. Outside the sleeping bag, the sun is already high into the sky. The air around is clear and dry.

I get up puffy-eyed and arthritic from the ground.

My mouth is already dry and cracked and my face and hands are covered with mosquito bites. Some sunburn from yesterday morning is hurting.

Beyond the pines are burned grass and clumps of earth and sand so bright they are hard to look at. The heat, silence, and barren hills and blank sky give a feeling of great, intense space.

Not a bit of moisture in the sky. Today's going to be a scorcher.

I walk out of the pines onto a stretch of barren sand between some grass and watch for a long time, meditatively. . . .

I've decided today's Chautauqua will begin to explore Phaedrus' world. It was intended earlier simply to restate some of his ideas that relate to technology and human values and make no reference to him personally, but the pattern of thought and memory that occurred last night has indicated this is not the way to go. To omit him now would be to run from something that should not be run from.

In the first grey of the morning what Chris said about his Indian friend's grandmother came back to me, clearing something up. She said ghosts appear when someone has not been buried right. That's true. He never *was* buried right, and that's exactly the source of the trouble.

Later I turn and see John is up and looking at me uncomprehendingly. He is still not really awake, and now walks aimlessly in circles to clear his head. Soon Sylvia is up too and her left eye is all puffed up. I ask her what happened. She says it is from mosquito bites. I begin to collect gear to repack the cycle. John does the same.

When this is done we get a fire started while Sylvia opens up packages of bacon and eggs and bread for breakfast.

When the food is ready, I go over and wake Chris. He doesn't want to get up. I tell him again. He says no. I grab the bottom of the sleeping bag, give it a mighty tablecloth jerk, and he is out of it, blinking in the pine needles. It takes him a while to figure out what has happened, while I roll up the sleeping bag.

He comes to breakfast looking insulted, eats one bite, says he isn't hungry, his stomach hurts. I point to the lake down below us, so strange in the middle of the semidesert, but he doesn't show any interest. He repeats his complaint. I just let it go by and John and

Sylvia disregard it too. I'm glad they were told what the situation is with him. It might have created real friction otherwise.

We finish breakfast silently, and I'm oddly tranquil. The decision about Phaedrus may have something to do with it. But we are also perhaps a hundred feet above the reservoir, looking across it into a kind of Western spaciousness. Barren hills, no one anywhere, not a sound; and there is something about places like this that raises your spirits a little and makes you think that things will probably get better.

While loading the remaining gear on the luggage rack I see with surprise that the rear tire is worn way down. All that speed and heavy load and heat on the road yesterday must have caused it. The chain is also sagging and I get out the tools to adjust it and then groan.

"What's the matter," John says.

"Thread's stripped in the chain adjustment."

I remove the adjusting bolt and examine the threads. "It's my own fault for trying to adjust it once without loosening the axle nut. The bolt is good." I show it to him. "It looks like the internal threading in the frame that's stripped."

John stares at the wheel for a long time. "Think you can make it into town?"

"Oh, yeah, sure. You can run it forever. It just makes the chain difficult to adjust."

He watches carefully as I take up the rear axle nut until it's barely snug, tap it sideways with a hammer until the chain slack is right, then tighten up the axle nut with all my might to keep the axle from slipping forward later on, and replace the cotter pin. Unlike the axle nuts on a car, this one doesn't affect bearing tightness.

"How did you know how to do that?" he asks.

"You just have to figure it out."

"I wouldn't know where to start," he says.

I think to myself, That's the problem, all right, where to start. To reach him you have to back up and back up, and the further back you go, the further back you see you have to go, until what looked like a small problem of communication turns into a major philosophic inquiry. That, I suppose, is why the Chautauqua.

I repack the tool kit and close the side cover plates and think to myself, He's worth reaching though.

On the road again the dry air cools off the slight sweat from that chain job and I'm feeling good for a while. As soon as the sweat dries off though, it's hot. Must be in the eighties already.

There's no traffic on this road, and we're moving right along. It's a traveling day.

Now I want to begin to fulfill a certain obligation by stating that there was one person, no longer here, who had something to say, and who said it, but whom no one believed or really understood. Forgotten. For reasons that will become apparent I'd prefer that he remain forgotten, but there's no choice other than to reopen his case.

I don't know his whole story. No one ever will, except Phaedrus himself, and he can no longer speak. But from his writings and from what others have said and from fragments of my own recall it should be possible to piece together some kind of approximation of what he was talking about. Since the basic ideas for this Chautauqua were taken from him there will be no real deviation, only an enlargement that may make the Chautauqua more understandable than if it were presented in a purely abstract way. The purpose of the enlargement is not to argue for him, certainly not to praise him. The purpose is to bury him—forever.

Back in Minnesota when we were traveling through some marshland I did some talking about the "shapes" of technology, the "death force" that the Sutherlands seem to be running from. I want to move now in the opposite direction from the Sutherlands, *toward* that force and into its center. In doing so we will be entering Phaedrus' world, the only world he ever knew, in which all understanding is in terms of underlying form.

The world of underlying form is an unusual object of discussion because it is actually a *mode* of discussion itself. You discuss things in terms of their immediate appearance or you discuss them in terms of their underlying form, and when you try to discuss these modes of discussion you get involved in what could be called a platform problem. You have no platform from which to discuss them other than the modes themselves.

Previously I was discussing his world of underlying form, or at least the aspect of it called technology, from an external view. Now I think it's right to talk about that world of underlying form from its own point of view. I want to talk about the underlying form of the world of underlying form itself.

To do this, first of all, a dichotomy is necessary, but before I can use it honestly I have to back up and say what *it* is and means, and that is a long story in itself. Part of this back-up problem. But right now I just want to use a dichotomy and explain it later. I want to divide human understanding into two kinds—classical understanding and romantic understanding. In terms of ultimate truth a dichotomy of this sort has little meaning but it is quite legitimate when one is operating within the classic mode used to discover or create a world of underlying form. The terms *classic* and *romantic,* as Phaedrus used them, mean the following:

A classical understanding sees the world primarily as underlying form itself. A romantic understanding sees it primarily in terms of immediate appearance. If you were to show an engine or a mechanical drawing or electronic schematic to a romantic it is unlikely he would see much of interest in it. It has no appeal because the reality he sees is its surface. Dull, complex lists of names, lines and numbers. Nothing interesting. But if you were to show the same blueprint or schematic or give the same description to a classical person he might look at it and then become fascinated by it because he sees that within the lines and shapes and symbols is a tremendous richness of underlying form.

The romantic mode is primarily inspirational, imaginative, creative, intuitive. Feelings rather than facts predominate. "Art" when it is opposed to "Science" is often romantic. It does not proceed by reason or by laws. It proceeds by feeling, intuition and esthetic conscience. In the northern European cultures the romantic mode is usually associated with femininity, but this is certainly not a necessary association.

The classic mode, by contrast, proceeds by reason and by laws—which are themselves underlying forms of thought and behavior. In the European cultures it is primarily a masculine mode and the fields of science, law and medicine are unattractive to women largely for this reason. Although motorcycle riding is romantic, motorcycle maintenance is purely classic. The dirt, the grease, the mastery of underlying form required all give it such a negative romantic appeal that women never go near it.

Although surface ugliness is often found in the classic mode of understanding it is not inherent in it. There is a classic esthetic which romantics often miss because of its subtlety. The classic style is straightforward, unadorned, unemotional, economical and carefully proportioned. Its purpose is not to inspire emotionally, but to bring order out of chaos and make the unknown known. It is not an esthetically free and natural style. It is esthetically restrained. Everything is under control. Its value is measured in terms of the skill with which this control is maintained.

To a romantic this classic mode often appears dull, awkward and ugly, like mechanical maintenance itself. Everything is in terms of pieces and parts and components and relationships. Nothing is figured out until it's run through the computer a dozen times. Everything's got to be measured and proved. Oppressive. Heavy. Endlessly grey. The death force.

Within the classic mode, however, the romantic has some appearances of his own. Frivolous, irrational, erratic, untrustworthy, interested primarily in pleasure-seeking. Shallow. Of no substance. Often a parasite who cannot or will not carry his own weight. A real drag on society. By now these battle lines should sound a little familiar.

This is the source of the trouble. Persons tend to think and feel exclusively in one mode or the other and in doing so tend to misunderstand and underestimate what the other mode is all about. But no one is willing to give up the truth as he sees it, and as far as I know, no one now living has any real reconciliation of these truths or modes. There is no point at which these visions of reality are unified.

And so in recent times we have seen a huge split develop between a classic culture and a romantic counterculture—two worlds growingly alienated and hateful toward each other with everyone wondering if it will always be this way, a house divided against itself. No one wants it really—despite what his antagonists in the other dimension might think.

It is within this context that what Phaedrus thought and said is significant. But *no* one was listening at that time and they only thought him eccentric at first, then undesirable, then slightly mad, and then genuinely insane. There seems little doubt that he was insane, but much of his writing at the time indicates that what was driving him insane was this hostile opinion of him. Unusual behavior tends to produce estrangement in others which tends to further the unusual behavior and thus the estrangement in self-stoking cycles until some sort of climax is reached. In Phaedrus' case there was a court-ordered police arrest and permanent removal from society.

I see we are at the left turn onto US 12 and John has pulled up for gas. I pull up beside him.

The thermometer by the door of the station reads 92 degrees. "Going to be another rough one today," I say.

When the tanks are filled we head across the street into a restaurant for coffee. Chris, of course, is hungry.

I tell him I've been waiting for that. I tell him he eats with the rest of us or not all. Not angrily. Just matter-of-factly. He's reproachful but sees how it's going to be.

I catch a fleeting look of relief from Sylvia. Evidently she thought this was going to be a continuous problem.

When we have finished the coffee and are outside again the heat is so ferocious we move off on the cycles as fast as possible. Again there is that momentary coolness, but it disappears. The sun makes the burned grass and sand so bright I have to squint to cut down glare. This US 12 is old, bad highway. The broken concrete is tar-patched and bumpy. Road signs indicate detours ahead. On either side of the road are occasional worn sheds and shacks and roadside stands that have accumulated through the years. The traffic is heavy now. I'm just as happy to be thinking about the rational, analytical, classical world of Phaedrus.

His kind of rationality has been used since antiquity to remove oneself from the tedium and depression of one's immediate surroundings. What makes it hard to see is that where once it was used to get away from it all, the escape has been so successful that now it is the "it all" that the romantics are trying to escape. What makes his world so hard to see clearly is not its strangeness but its usualness. Familiarity can blind you too.

His way of looking at things produces a kind of description that can be called an "analytic" description. That is another name of the classic platform from which one discusses things in terms of their underlying form. He was a totally classic person. And to give a fuller description of what this is I want now to turn his analytic approach back upon itself—to analyze analysis itself. I want to do this first of all by giving an extensive example of it and then by dissecting what it is. The motorcycle is a perfect subject for it since the motorcycle itself was invented by classic minds. So listen:

A motorcycle may be divided for purposes of classical rational analysis by means of its component assemblies and by means of its functions.

If divided by means of its component assemblies, its most basic division is into a power assembly and a running assembly.

The power assembly may be divided into the engine and the power-delivery system. The engine will be taken up first.

The engine consists of a housing containing a power train, a fuel-air system, an ignition system, a feedback system and a lubrication system.

The power train consists of cylinders, pistons, connecting rods, a crankshaft and a flywheel.

The fuel-air system components, which are part of the engine, consist of a gas tank and filter, an air cleaner, a carburetor, valves and exhaust pipes.

The ignition system consists of an alternator, a rectifier, a battery, a high-voltage coil and spark plugs.

The feedback system consists of a cam chain, a camshaft, tappets and a distributor.

The lubrication system consists of an oil pump and channels throughout the housing for distribution of the oil.

The power-delivery system accompanying the engine consists of a clutch, a transmission and a chain.

The supporting assembly accompanying the power assembly consists of a frame, including foot pegs, seat and fenders; a steering assembly; front and rear shock absorbers; wheels; control levers and cables; lights and horn; and speed and mileage indicators.

That's a motorcycle divided according to its components. To know what the components are for, a division according to functions is necessary:

A motorcycle may be divided into normal running functions and special, operator-controlled functions.

Normal running functions may be divided into functions during the intake cycle, functions during the compression cycle, functions during the power cycle and functions during the exhaust cycle.

And so on. I could go on about which functions occur in their proper sequence during each of the four cycles, then go on to the operator-controlled functions and that would be a very summary description of the underlying form of a motorcycle. It would be extremely short and rudimentary, as descriptions of this sort go. Almost any one of the components mentioned can be expanded on indefinitely. I've read an entire engineering volume on contact points alone, which are just a small but vital part of the distributor. There are other types of engines than the single-cylinder Otto engine described here: two-cycle engines, multiple-cylinder engines, diesel engines, Wankel engines—but this example is enough.

This description would cover the "what" of the motorcycle in terms of components, and the "how" of the engine in terms of functions. It would badly need a "where" analysis in the form of an illustration, and also a "why" analysis in the form of engineering principles that led to this particular conformation of parts. But the purpose here isn't exhaustively to analyze the motorcycle. It's to provide a starting point, an example of a mode of understanding of things which will itself become an object of analysis.

There's certainly nothing strange about this description at first hearing. It sounds like something from a beginning textbook on the subject, or perhaps a first lesson in a vocational course. What is unusual about it is seen when it ceases to be a mode of discourse and becomes an object of discourse. Then certain things can be pointed to.

The first thing to be observed about this description is so obvious you have to hold it down or it will drown out every other observation. That is: It is just duller than ditchwater. Yah-da, yah-da, yah-da, yah-da, yah, carburetor, gear ratio, compression, yah-da-yah, piston, plugs, intake, yah-da-yah, on and on and on. That is the romantic face of the classic mode. Dull, awkward and ugly. Few romantics get beyond that point.

But if you can hold down that most obvious observation, some other things can be noticed that do not at first appear.

The first is that the motorcycle, so described, is almost impossible to understand unless you already know how one works. The immediate surface impressions that are essential for primary understanding are gone. Only the underlying form is left.

The second is that the observer is missing. The description doesn't say that to see the piston you must remove the cylinder head. "You" aren't anywhere in the picture. Even the "operator" is a kind of personality-less robot whose performance of a function on the machine is completely mechanical. There are no real subjects in this description. Only objects exist that are independent of any observer.

The third is that the words *"good"* and *"bad"* and all their synonyms are completely absent. No value judgments have been expressed anywhere, only facts.

The fourth is that there is a knife moving here. A very deadly one; an intellectual scalpel so swift and so sharp you sometimes don't see it moving. You get the illusion that all those parts are just there and are being named as they exist. But they can be named quite differently and organized quite differently depending on how the knife moves.

For example, the feedback mechanism which includes the camshaft and cam chain and tappets and distributor exists only because of an unusual cut of this analytic knife. If you were to go to a motorcycle-parts department and ask them for a feedback assembly they wouldn't know what the hell you were talking about. They don't split it up that way. No two manufacturers ever split it up quite the same way and every mechanic is familiar with the problem of the part you can't buy because you can't find it because the manufacturer considers it a part of something else.

It is important to see this knife for what it is and not to be fooled into thinking that motorcycles or anything else are the way they are just because the knife happened to cut it up that way. It is important to concentrate on the knife itself. Later I will want to show how an ability to use this knife creatively and effectively can result in solutions to the classic and romantic split.

Phaedrus was a master with this knife, and used it with dexterity and a sense of power. With a single stroke of analytic thought he split the whole world into parts of his own choosing, split the parts and split the fragments of the parts, finer and finer and finer until he had reduced it to what he wanted it to be. Even the special use of the terms *"classic"* and *"romantic"* are examples of his knifemanship.

But if this were all there were to him, analytic skill, I would be more than willing to shut up about him. What makes it important not to shut up about him was that he used this skill in such a bizarre and yet meaningful way. No one ever saw this, I don't think he even saw it himself, and it may be an illusion of my own, but the knife he used was less that of an assassin than that of a poor surgeon. Perhaps there is no difference. But he saw a sick and ailing thing happening and he started cutting deep, deeper and deeper to get at the root of it. He was after something. That is important. He was after something and he used the knife because that was the only tool he had. But he took on so much and went so far in the end his real victim was himself.

7

Heat is everywhere now. I can't ignore it anymore. The air is like a furnace blast so hot that my eyes under the goggles feel cool compared to the rest of my face. My hands are cool but the gloves have big black spots from perspiration on the back surrounded by white streaks of dried salt.

On the road ahead a crow tugs on some carrion and flies up slowly as we approach. It looks like a lizard on the road, dry and stuck to the tar.

On the horizon appears an image of buildings, shimmering slightly. I look down at the map and figure it must be Bowman. I think about ice water and air conditioning.

On the street and sidewalks of Bowman we see almost no one, even though plenty of parked cars show they're here. All inside. We swing the machines into an angled parking place with a tight turn that points them outward, for when we're ready to go. A lone, elderly person wearing a broad-brimmed hat watches us put the cycles on their stands and remove helmets and goggles.

"Hot enough for you?" he asks. His expression is blank.

John shakes his head and says, "Gawd!"

The expression, shaded by the hat, becomes almost a smile.

"What *is* the temperature?" John asks.

"Hundred and two," he says, "last I saw. Should go to hundred and four."

He asks us how far we have come and we tell him and he nods with a kind of approval. "That's a long way," he says. Then he asks about the machines.

The beer and air conditioning are calling, but we don't break away. We just stand there in the hundred-and-two sun talking to this person. He is a stockman, retired, says this is pretty much ranch country around here and he used to own a Henderson cycle years ago. It pleases me that he should want to talk about his Henderson in this hundred-and-two sun. We talk about it for a while, with growing impatience from John and Sylvia and Chris, and when we finally say good-bye he says he is glad to have met us and his expression is still blank but we sense that he really meant it. He walks away with a kind of slow dignity in the hundred-and-two sun.

In the restaurant I try to comment on this but no one is interested. John and Sylvia look really out of it. They just sit and soak up the air-conditioned air without a move. The waitress comes for the order and that snaps them out of it a little, but they are not ready and so she goes away again.

"I don't think I want to leave here," Sylvia says.

An image of the elderly man outside in the wide-brimmed hat comes back to me. "Think what it was like around here before air conditioning," I say.

"I am," she says.

"With the roads this hot and that bad back tire of mine, we shouldn't go more than sixty," I say.

No comment from them.

Chris, in contrast to them, seems to be back to his normal self, alert and watching everything. When the food comes he wolfs it down and then, before we are half-finished, asks for more. He gets it and we wait for him to finish.

Miles later and the heat is just ferocious. Sunglasses and goggles are not enough for this glare. You need a welder's mask.

The High Plains break up into washed-out and gullied hills. It is all bright whitish tan. Not a blade of grass anywhere. Just scattered weed stalks and rocks and sand. The black of the highway is a relief to look at so I stare down at it and study how the blur whizzes by underfoot. Beside it I see the left exhaust pipe has picked up a bluer color than it has ever had before. I spit on my glove tips, touch it and can see the sizzle. Not good.

It's important now to just live with this and not fight it mentally . . . mind control. . . .

I should talk now about Phaedrus' knife. It'll help understand some of the things we talked about.

The application of this knife, the division of the world into parts and the building of this structure, is something everybody does. All the time we are aware of millions of things around us—these changing shapes, these burning hills, the sound of the engine, the feel of the throttle, each rock and weed and fence post and piece of debris beside the road—aware of these things but not really conscious of them unless there is something unusual or unless they reflect something we are predisposed to see. We could not possibly be conscious of these things and remember all of them because our mind would be so full of useless details we would be unable to think. From all this awareness we must select, and what we select and call consciousness is never the same as the awareness because the process of selection mutates it. We take a handful of sand from the endless landscape of awareness around us and call that handful of sand the world.

Once we have the handful of sand, the world of which we are conscious, a process of discrimination goes to work on it. This is the knife. We divide the sand into parts. This and that. Here and there. Black and white. Now and then. The discrimination is the division of the conscious universe into parts.

The handful of sand looks uniform at first, but the longer we look at it the more diverse we find it to be. Each grain of sand is different. No two are alike. Some are similar in one way, some are similar in another way, and we can form the sand into separate piles on the basis of this similarity and dissimilarity. Shades of color in different piles—sizes in different piles—grain shapes in different piles—subtypes of grain shapes in different piles—grades of opacity in different piles—and so on, and on, and on. You'd think the process of subdivision and classification would come to an end somewhere, but it doesn't. It just goes on and on.

Classical understanding is concerned with the piles

and the basis for sorting and interrelating them. Romantic understanding is directed toward the handful of sand before the sorting begins. Both are valid ways of looking at the world although irreconcilable with each other.

What has become an urgent necessity is a way of looking at the world that does violence to neither of these two kinds of understanding and unites them into one. Such an understanding will not reject sand-sorting *or* contemplation of unsorted sand for its own sake. Such an understanding will instead seek to direct attention to the endless landscape from which the sand is taken. That is what Phaedrus, the poor surgeon, was trying to do.

To understand what he was trying to do it's necessary to see that *part* of the landscape, *inseparable* from it, which *must* be understood, is a figure in the middle of it, sorting sand into piles. To see the landscape without seeing this figure is not to see the landscape at all. To reject that part of the Buddha that attends to the analysis of motorcycles is to miss the Buddha entirely.

There is a perennial classical question that asks which part of the motorcycle, which grain of sand in which pile, is the Buddha. Obviously to ask that question is to look in the wrong direction, for the Buddha is everywhere. But just as obviously to ask that question is to look in the *right* direction, for the Buddha is everywhere. About the Buddha that exists independently of any analytic thought much has been said—some would say *too* much, and would question any attempt to add to it. But about the Buddha that exists *within* analytic thought, and *gives that analytic thought its direction,* virtually nothing has been said, and there are historic reasons for this. But history keeps happening, and it seems no harm and maybe some positive good to add to our historical heritage with some talk in this area of discourse.

When analytic thought, the knife, is applied to experience, something is always killed in the process. That is fairly well understood, at least in the arts. Mark Twain's experience comes to mind, in which, after he had mastered the analytic knowledge needed to pilot the Mississippi River, he discovered the river had lost its beauty. Something *is* always killed. But what is less noticed in the arts—something is always created too. And instead of just dwelling on what is killed it's important also to see what's created and to see the process as a kind of death-birth continuity that is neither good nor bad, but just *is*.

We pass through a town called Marmarth but John doesn't stop even for a rest and so we go on. More furnace heat, into some badlands, and we cross the border into Montana. A sign by the road announces it.

Sylvia waves her arms up and down and I beep the horn in response, but when I look at the sign my feelings are not jubilant at all. For me its information causes a sudden inward tension that can't exist for them. They've no way of knowing we're now in the country where he lived.

All this talk so far about classic and romantic understanding must seem a strangely oblique way of describing him, but to get at Phaedrus, this oblique route is the only one to take. To describe his physical appearance or the statistics of his life would be to dwell on misleading superficialities. And to come at him directly would be to invite disaster.

He was insane. And when you look directly at an insane man all you see is a reflection of your own knowledge that he's insane, which is not to see him at all. To see him you must see what he saw and when you are trying to see the vision of an insane man, an oblique route is the only way to come at it. Otherwise your own opinions block the way. There is only one access to him that I can see as passable and we still have a way to go.

I've been going into all this business of analyses and definitions and hierarchies not for their own sake but to lay the groundwork for an understanding of the direction in which Phaedrus went.

I told Chris the other night that Phaedrus spent his entire life pursuing a ghost. That was true. The ghost he pursued was the ghost that underlies all of technology, all of modern science, all of Western thought. It was the ghost of rationality itself. I told Chris that he found the ghost and that when he found it he thrashed it good. I think in a figurative sense that is true. The things I hope to bring to light as we go along are some of the things he uncovered. Now the times are such that others may at last find them of value. No one then would see the ghost that Phaedrus pursued, but I think now that more and more people see it, or get glimpses of it in bad moments, a ghost which calls itself rationality but whose appearance is that of incoherence and meaninglessness, which causes the most normal of everyday acts to seem slightly mad because of their irrelevance to anything else. This is the ghost of normal everyday assumptions which declares that the ultimate purpose of life, which is to keep alive, is impossible, but that this is the ultimate purpose of life anyway, so that great minds struggle to cure diseases so that people may live longer, but only madmen ask why. One lives longer in order that he may live longer. There is no other purpose. That is what the ghost says.

At Baker, where we stop, the thermometers are reading 108 degrees in the shade. When I take my gloves off, the metal of the gas tank is so hot I can't touch it. The engine is making ominous knick-knicking sounds from overheating. Very bad. The rear tire has worn badly too, and I feel with my hand that it's almost as hot as the gas tank.

"We're going to have to slow down," I say.

"What?"

"I don't think we should go over fifty," I say.

John looks at Sylvia and she looks at him. Something has already been said between them about my slowness. They both look as if they've about had it.

"We just want to get there fast," John says, and they both walk toward a restaurant.

The chain has been running hot and dry too. In the right-hand saddlebag I rummage for a can of spray lubricant, find it, then start the engine and spray the moving chain. The chain is still so hot the solvent evaporates almost instantly. Then I squirt a little oil on, let it run for a minute and shut the engine off. Chris waits patiently, then follows me into the restaurant.

"I thought you said the big slump was going to come on the second day," Sylvia says as we approach the booth they are in.

"Second or third," I reply.

"Or fourth or fifth?"

"Maybe."

She and John look at each other again with the same expression they showed before. It seems to say, "Three's a crowd." They may want to go ahead fast and wait for me in some town up ahead. I'd suggest it myself except that if they go much faster they won't be waiting for me in some town. It'll be by the side of the road.

"I don't know how the people here stand this," Sylvia says.

"Well, it's hard country," I say with a little irritation. "They know it's hard before they come here and are ready for it."

I add, "If one person complains he just makes it that much harder for the others. They've got *stamina*. They know how to keep on going."

John and Sylvia don't say much, and John finishes his Coke early and is off to a bar for a snort. I go out and check the cycle luggage again and find that the new pack has been compressing a little and so take up the slack in the ropes and retie them.

Chris points to a thermometer in direct sunlight and we see it has gone all the way above the scale at 120 degrees.

Before we are out of town I am sweating again. The cool drying-off period doesn't last even half a minute.

The heat just slams into us. Even with dark sunglasses I have to squint my eyes into slits. There's nothing but burning sand and pale sky so bright it's hard to look anywhere. It's just become white-hot everywhere. A real inferno.

John up ahead is speeding faster and faster. I give up on him and slow it down to fifty-five. Unless you're just looking for trouble in this heat you don't run tires at eighty-five. A blowout on this stretch would really be it.

I suppose they took what I said as a kind of rebuke but I didn't have that in mind. I'm no more comfortable than they are in this heat but there's no point in dwelling on it. All day while I've been thinking and talking about Phaedrus they must have been thinking about how bad all this is. That's what's really wearing them down. The thought.

Some things can be said about Phaedrus as an individual:

He was a knower of logic, the classical system-of-the-system which describes the rules and procedures of systematic thought by which analytic knowledge may be structured and interrelated. He was so swift at this his Stanford-Binet IQ, which is essentially a record of skill at analytic manipulation, was recorded at 170, a figure that occurs in only one person in fifty thousand.

He was systematic, but to say he thought and acted like a machine would be to misunderstand the nature of his thought. It was not like pistons and wheels and gears all moving at once, massive and coordinated. The image of a laser beam comes to mind instead; a single pencil of light of such terrific energy in such extreme concentration it can be shot at the moon and its reflection seen back on earth. Phaedrus did not try to use his brilliance for general illumination. He sought one specific distant target and aimed for it and hit it. And that was all. General illumination of that target he hit now seems to be left for me.

In proportion to his intelligence he was extremely isolated. There's no record of his having had close friends. He traveled alone. Always. Even in the presence of others he was completely alone. People sometimes felt this and felt rejected by it, and so did not like him, but their dislike was not important to him.

His wife and family seem to have suffered the most. His wife says those who tried to go beyond the barriers of his reserve found themselves facing a blank. My impression is that they were starved for some kind of affection which he never gave.

No one really knew him. That is evidently the way he wanted it, and that's the way it was. Perhaps his aloneness was the result of his intelligence. Perhaps it was the cause. But the two were always together. An uncanny solitary intelligence.

This still doesn't do it though, because this and the image of a laser beam convey the idea that he was completely cold and unemotional, and that is not so. In his pursuit of what I have called the ghost of rationality he was a fanatic hunter.

One fragment becomes especially vivid now of a scene in the mountains where the sun was behind the mountain half an hour and an early twilight had changed the trees and even the rocks to almost blackened shades of blue and grey and brown. Phaedrus had been there three days without food. His food had run out but he was thinking deeply and seeing things and was reluctant to leave. He was not far away from where he knew there was a road and was in no hurry.

In the dusk coming down the trail he saw a movement and then what seemed to be a dog approaching on the trail, a very large sheep dog, or an animal more like a husky, and he wondered what would bring a dog to this obscure place at this time of evening. He disliked dogs, but this animal moved in a way that fore-stalled these feelings. It seemed to be watching him, judging *him*. Phaedrus stared into the animal's eyes for a long time, and for a moment felt some kind of recognition. Then the dog disappeared.

He realized much later it was a timber wolf, and the memory of this incident stayed with him a long time. I think it stayed with him because he had seen a kind of image of himself.

A photograph can show a physical image in which time is static, and a mirror can show a physical image in which time is dynamic, but I think what he saw on the mountain was another kind of image altogether which was not physical and did not exist in time at all. It was an image nevertheless and that is why he felt recognition. It comes to me vividly now because I saw it again last night as the visage of Phaedrus himself.

Like that timber wolf on the mountain he had a kind of animal courage. He went his own way with uncon-cern for consequences that sometimes stunned people, and stuns me now to hear about it. He did not often swerve to right or to left. I've discovered that. But this courage didn't arise from any idealistic idea of self-sacrifice, only from the intensity of his pursuit, and there was nothing noble about it.

I think his pursuit of the ghost of rationality oc-curred because he wanted to wreak *revenge* on it, be-cause he felt he himself was so shaped by it. He wanted to free himself from his own image. He wanted to destroy it because the ghost was what *he* was and he wanted to be free from the bondage of his own identity. In a strange way, this freedom was achieved.

This account of him must sound unworldly, but the

most unworldly part of it all is yet to come. This is my
own relationship to him. This has been forestalled and
obscured until now, but nevertheless must be known.

I first discovered him by inference from a strange
series of events many years ago. One Friday I had gone
to work and gotten quite a lot done before the weekend
and was happy about that and later that day drove to a
party where, after talking to everybody too long and
too loudly and drinking *way* too much, went into a
back room to lie down for a while.

When I awoke I saw that I'd slept the whole night,
because now it was daylight, and I thought, "My God,
I don't even know the name of the hosts!" and wondered what kind of embarrassment this was going to
lead to. The room didn't look like the room I had lain
down in, but it had been dark when I came in and I
must have been blind drunk anyway.

I got up and saw that my clothes were changed.
These were not the clothes I had worn the night before.
I walked out the door, but to my surprise the doorway
led not to rooms of a house but into a long corridor.

As I walked down the corridor I got the impression
that everyone was looking at me. Three different times
a stranger stopped me and asked how I felt. Thinking
they were referring to my drunken condition I replied
that I didn't even have a hangover, which caused one
of them to start to laugh, but then catch himself.

At a room at the end of the corridor I saw a table
where there was activity of some sort going on. I sat
down nearby, hoping to remain unnoticed until I got all
this figured out. But a woman dressed in white came
up to me and asked if I knew her name. I read the little name clip on her blouse. She didn't see that I was

doing this and seemed amazed, and walked off in a hurry.

When she came back there was a man with her, and he was looking right at me. He sat down next to me and asked me if I knew his name. I told him what it was, and was as surprised as they were that I knew it.

"It's very early for this to be happening," he said.

"This looks like a hospital," I said.

They agreed.

"How did I get here?" I asked, thinking about the drunken party. The man said nothing and the woman looked down. Very little was explained.

It took me more than a week to deduce from the evidence around me that everything before my waking up was a dream and everything afterward was reality. There was no basis for distinguishing the two other than the growing pile of new events that seemed to argue against the drunk experience. Little things appeared, like the locked door, the outside of which I could never remember seeing. And a slip of paper from the probate court telling me that some person was committed as insane. Did they mean *me*?

It was explained to me finally that "You have a new personality now." But this statement was no explanation at all. It puzzled me more than ever since I had no awareness at all of any "old" personality. If they had said, "You *are* a new personality," it would have been much clearer. That would have fitted. They had made the mistake of thinking of a personality as some sort of possession, like a suit of clothes, which a person wears. But apart from a personality what is there? Some bones and flesh. A collection of legal statistics, perhaps, but surely no person. The bones and flesh and legal statistics are the garments worn by the personality, not the other way around.

But who was the *old* personality whom they had known and presumed I was a continuation of?

This was my first inkling of the existence of Phaedrus, many years ago. In the days and weeks and years that have followed, I've learned much more.

He was dead. Destroyed by order of the court, enforced by the transmission of high-voltage alternating current through the lobes of his brain. Approximately 800 mills of amperage at durations of 0.5 to 1.5 seconds had been applied on twenty-eight consecutive occasions, in a process known technologically as "Annihilation ECS." A whole personality had been liquidated without a trace in a technologically faultless act that has defined our relationship ever since. I have never met him. Never will.

And yet strange wisps of his memory suddenly match and fit this road and desert bluffs and white-hot sand all around us and there is a bizarre concurrence and then I know he has seen all of this. He was here, otherwise I would not know it. He had to be. And in seeing these sudden coalescences of vision and in recall of some strange fragment of thought whose origin I have no idea of, I'm like a clairvoyant, a spirit medium receiving messages from another world. That is how it is. I see things with my own eyes, and I see things with his eyes too. He once owned them.

These EYES! That is the terror of it. These gloved hands I now look at, steering the motorcycle down the road, were once *his*! And if you can understand the feeling that comes from that, then you can understand real fear—the fear that comes from knowing there is nowhere you can possibly run.

We enter a low-rimmed canyon. Before long, a roadside stop I've been waiting for appears. A few benches, a little building and some tiny green trees with hoses running to their bases. John, so help me God, is at the exit on the other side, ready to pull out onto the highway.

I ignore this and pull up by the building. Chris jumps off and we pull the machine back up on the stand. The heat rises from the engine as if it were on fire, throwing off waves that distort everything around it. Out of the corner of my eye I see the other cycle come back. When they arrive they are both glaring at me.

Sylvia says, "We're just . . . angry!"

I shrug my shoulders and walk to the drinking fountain.

John says, "Where's all that stamina you were telling us about!"

I look at him for a second and see he really *is* angry. "I was afraid you took that too seriously," I say, and then turn away. I drink the water and it's alkaline, like soapy water. I drink it anyway.

John goes into the building to soak his shirt with water. I check the oil level. The oil filter cap is so hot it burns my fingers right through the gloves. The engine hasn't lost much oil. The back tire tread is down a little more but still serviceable. The chain is tight enough but a little dry so I oil it again to be safe. The critical bolts are all tight enough.

John comes over dripping with water and says, "You go ahead this time, we'll stay behind."

"I won't go fast," I say.

"That's all right," he says. "We'll get there."

So I go ahead and we take it slowly. The road through the canyon doesn't straighten out into more of what we've been through, as I expected it would, but starts to wind upward. Surprise.

Now the road meanders a little, now it cuts back away from the direction in which we should be going, then returns. Soon it rises a little and then rises some more. We are moving in angular directions into narrow devil's gaps, then upward again higher and a little higher each time.

Some shrubs appear. Then small trees. The road goes higher still into grass, and then fenced meadows.

Overhead a small cloud appears. Rain perhaps? Perhaps. Meadows must have rain. And these now have flowers in them. Strange how all this has changed. Nothing to show it on the map. And the consciousness of memory has disappeared too. Phaedrus must not have come this way. But there was no other road. Strange. It keeps rising upward.

The sun angles toward the cloud, which now has grown downward to touch the horizon above us, in which there are trees, pines, and a cold wind comes down with pine smells from the trees. The flowers in the meadow blow in the wind and the cycle leans a little and we are suddenly cool.

I look at Chris and he is smiling. I am smiling too.

Then the rain comes hard on the road with a gust of earth-smell from the dust that has waited for too long and the dust beside the road is pocked with the first raindrops.

This is all so new. And we are so in need of it, a new rain. My clothes become wet, and goggles are spattered, and chills start and feel delicious. The cloud passes from beneath the sun and the forest of pines and small meadows gleams again, sparkling where the sunlight catches small drops from the rain.

We reach the top of the climb dry again but cool now and stop, overlooking a huge valley and river below.

"I think we have arrived," John says.

Sylvia and Chris have walked into the meadow among the flowers under pines through which I can see the far side of the valley, away and below.

I am a pioneer now, looking onto a promised land.

PART II

8

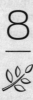

It's about ten o'clock in the morning and I'm sitting alongside the machine on a cool, shady curbstone back of a hotel we have found in Miles City, Montana. Sylvia is with Chris at a Laundromat doing the laundry for all of us. John is off looking for a duckbill to put on his helmet. He thought he saw one at a cycle shop when we came into town yesterday. And I'm about to sharpen up the engine a little.

Feeling good now. We got in here in the afternoon and made up for a lot of sleep. It was a good thing we stopped. We were so stupid with exhaustion we didn't know how tired we were. When John tried to register rooms he couldn't even remember my name. The desk girl asked us if we owned those "groovy, dreamy motorcycles" outside the window and we both laughed so hard she wondered what she had said wrong. It was just numbskull laughter from too much fatigue. We've been more than glad to leave them parked and walk for a change.

And baths. In a beautiful old enameled cast-iron bathtub that crouched on lion's paws in the middle of a marble floor, just waiting for us. The water was so soft it felt as if I would never get the soap off. Afterward we walked up and down the main streets and felt like a family. . . .

On this machine I've done the tuning so many times it's become a ritual. I don't have to think much about how to do it anymore. Just mainly look for anything unusual. The engine has picked up a noise that sounds like a loose tappet but could be something worse, so I'm going to tune it now and see if it goes away. Tappet adjustment has to be done with the engine cold, which means wherever you park it for the night is where you work on it the next morning, which is why I'm on a shady curbstone back of a hotel in Miles City, Montana. Right now the air is cool in the shade and will be for an hour or so until the sun gets around the tree branches, which is good for working on cycles. It's important not to tune these machines in the direct sun or late in the day when your brain gets muddy because even if you've been through it a hundred times you should be alert and looking for things.

Not everyone understands what a completely rational process this is, this maintenance of a motorcycle. They think it's some kind of a "knack" or some kind of "affinity for machines" in operation. They are right, but the knack is almost purely a process of reason, and most of the troubles are caused by what old time radio men called a "short between the earphones," failures to use the head properly. A motorcycle functions entirely in accordance with the laws of reason, and a study of the art of motorcycle maintenance is really a miniature study of the art of rationality itself. I said yesterday that the ghost of rationality was what Phaedrus pursued and what led to his insanity, but to get into that it's vital to stay with down-to-earth examples of rationality, so as not to get lost in generalities no one else can understand. Talk about rationality can get very confusing unless the things with which rationality deals are also included.

We are at the classic-romantic barrier now, where on one side we see a cycle as it appears immediately—and this is an important way of seeing it—and where on the other side we can begin to see it as a mechanic does in terms of underlying form—and this is an important way of seeing things too. These tools for example—this wrench—has a certain romantic beauty to it, but its purpose is always purely classical. It's designed to change the underlying form of the machine.

The porcelain inside this first plug is very dark. That is classically as well as romantically ugly because it means the cylinder is getting too much gas and not enough air. The carbon molecules in the gasoline aren't finding enough oxygen to combine with and they're just sitting here loading up the plug. Coming into town yesterday the idle was loping a little, which is a symptom of the same thing.

Just to see if it's just the one cylinder that's rich I check the other one. They're both the same. I get out a pocket knife, grab a stick lying in the gutter and whittle down the end to clean out the plugs, wondering what could be the cause of the richness. That wouldn't have anything to do with rods or valves. And carbs rarely go out of adjustment. The main jets are over-sized, which causes richness at high speeds but the plugs were a lot cleaner than this before with the *same* jets. Mystery. You're always surrounded by them. But if you tried to solve them all, you'd never get the machine fixed. There's no immediate answer so I just leave it as a hanging question.

The first tappet is right on, no adjustment required, so I move on to the next. Still plenty of time before the sun gets past those trees . . . I always feel like I'm in church when I do this . . . The gauge is some kind of religious icon and I'm performing a holy rite with it. It is a member of a set called "precision measuring instruments" which in a classic sense has a profound meaning.

In a motorcycle this precision isn't maintained for any romantic or perfectionist reasons. It's simply that the enormous forces of heat and explosive pressure inside this engine can only be controlled through the kind of precision these instruments give. When each explosion takes place it drives a connecting rod onto the crankshaft with a surface pressure of many tons per square inch. If the fit of the rod to the crankshaft is precise the explosion force will be transferred smoothly and the metal will be able to stand it. But if the fit is loose by a distance of only a few thousandths of an inch the force will be delivered suddenly, like a hammer blow, and the rod, bearing and crankshaft surface will soon be pounded flat, creating a noise which at first sounds a lot like loose tappets. That's the reason I'm checking it now. If it *is* a loose rod and I try to make it to the mountains without an overhaul, it will soon get louder and louder until the rod tears itself free, slams into the spinning crankshaft and destroys the engine. Sometimes broken rods will pile right down through the crankcase and dump all the oil onto the road. All you can do then is start walking.

But all this can be prevented by a few thousandths of an inch fit which precision measuring instruments give, and this is their classical beauty—not what you see, but what they mean—what they are capable of in terms of control of underlying form.

The second tappet's fine. I swing over to the street side of the machine and start on the other cylinder.

Precision instruments are designed to achieve an *idea*, dimensional precision, whose perfection is impossible. There is no perfectly shaped part of the motorcycle and never will be, but when you come as close as these instruments take you, remarkable things happen, and you go flying across the countryside under a power that would be called magic if it were not so completely rational in every way. It's the understanding of this rational intellectual *idea* that's fundamental. John looks at the motorcycle and he sees steel in various shapes and has negative feelings about these steel shapes and turns off the whole thing. I look at the shapes of the steel now and I see *ideas*. He thinks I'm working on *parts*. I'm working on *concepts*.

I was talking about these concepts yesterday when I said that a motorcycle can be divided according to its components and according to its functions. When I said that suddenly I created a set of boxes with the following arrangement:

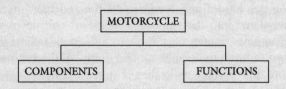

And when I said the components may be subdivided into a power assembly and a running assembly, suddenly appear some more little boxes:

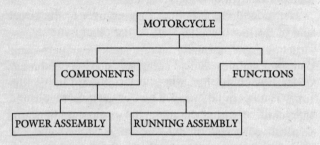

And you see that every time I made a further division, up came more boxes based on these divisions until I had a huge pyramid of boxes. Finally you see that while I was splitting the cycle up into finer and finer pieces, I was also building a structure.

This structure of concepts is formally called a hierarchy and since ancient times has been a basic structure for all Western knowledge. Kingdoms, empires, churches, armies have all been structured into hierarchies. Modern businesses are so structured. Tables of contents of reference material are so structured, mechanical assemblies, computer software, all scientific and technical knowledge is so structured—so much so that in some fields such as biology, the hierarchy of kingdom-phylum-class-order-family-genus-species is almost an icon.

The box "motorcycle" *contains* the boxes "components" and "functions." The box "components" *contains* the boxes "power assembly" and "running assembly," and so on. There are many other kinds of structures produced by other operators such as "causes" which produce long chain structures of the form, "A causes B which causes C which causes D," and so on. A functional description of the motorcycle uses this structure. The operators "exists," "equals," and "implies" produce still other structures. These structures are normally interrelated in patterns and paths so complex and so enormous no one person can understand more than a small part of them in his lifetime. The overall name of these interrelated structures, the genus of which the hierarchy of containment and structure of causation are just species, is *system.* The motorcycle is a system. A *real* system.

To speak of certain government and establishment institutions as "the system" is to speak correctly, since these organizations are founded upon the same structural conceptual relationships as a motorcycle. They are sustained by structural relationships even when they have lost all other meaning and purpose. People arrive at a factory and perform a totally meaningless task from eight to five without question because the structure demands that it be that way. There's no villain, no "mean guy" who wants them to live meaningless lives, it's just that the structure, the system demands it and no one is willing to take on the formidable task of changing the structure just because it is meaningless.

But to tear down a factory or to revolt against a government or to avoid repair of a motorcycle because it is a system is to attack effects rather than causes; and as long as the attack is upon effects only, no change is possible. The true system, the real system, is our present construction of systematic thought itself, rationality itself, and if a factory is torn down but the rationality which produced it is left standing, then that rationality will simply produce another factory. If a revolution destroys a systematic government, but the systematic patterns of thought that produced that government are left intact, then those patterns will repeat themselves in the succeeding government. There's so much talk about the system. And so little understanding.

That's all the motorcycle is, a system of concepts worked out in steel. There's no part in it, no shape in it, that is not out of someone's mind . . . number three tappet is right on too. One more to go. This had better be it. . . . I've noticed that people who have never worked with steel have trouble seeing this—that the motorcycle is primarily a mental phenomenon. They associate metal with given shapes—pipes, rods, girders, tools, parts—all of them fixed and inviolable, and think of it as primarily physical. But a person who does machining or foundry work or forge work or welding sees "steel" as having no shape at all. Steel can be any shape you want if you are skilled enough, and any shape *but* the one you want if you are not. Shapes, like this tappet, are what you *arrive* at, what you give to the steel. Steel has no more shape than this old pile of dirt on the engine here. These shapes are all out of someone's mind. That's important to see. The *steel*? Hell, even the steel is out of someone's mind. There's no steel in nature. Anyone from the Bronze Age could have told you that. All nature has is a *potential* for steel. There's nothing else there. But what's "potential"? That's also in someone's mind! . . . Ghosts.

That's really what Phaedrus was talking about when he said it's all in the mind. It sounds insane when you just jump up and say it without reference to anything specific like an engine. But when you tie it down to something specific and concrete, the insane sound tends to disappear and you see he could have been saying something of importance.

The fourth tappet *is* too loose, which is what I had hoped. I adjust it. I check the timing and see that it is still right on and the points are not pitted, so I leave them alone, screw on the valve covers, replace the plugs and start it up.

The tappet noise is gone, but that doesn't mean much yet while the oil is still cold. I let it idle while I pack the tools away, then climb on and head for a cycle shop a cyclist on the street told us about last night where they may have a chain adjuster link, and a new foot-peg rubber. Chris must have nervous feet. His foot pegs keep wearing out.

I go a couple of blocks and still no tappet noise. It's beginning to sound good, I think it's gone. I won't come to any conclusions until we've gone about thirty miles though. But until then, and right now, the sun is bright, the air is cool, my head is clear, there's a whole day ahead of us, we're almost to the mountains, it's a good day to be alive. It's this thinner air that does it. You always feel like this when you start getting into higher altitudes.

The altitude! That's why the engine's running rich. Sure, that's got to be the reason. We're at twenty-five hundred feet now. I'd better switch to standard jets. They take only a few minutes to put in. And lean out the idle adjustment a little. We'll be getting up a lot higher than this.

Under some shady trees I find Bill's Cycle Shop but no Bill. A passerby says he has "maybe gone fishing somewhere," leaving his shop wide open. We really *are* in the West. No one would leave a shop like this open in Chicago or New York.

Inside I see that Bill is a mechanic of the "photographic mind" school. Everything lying around everywhere. Wrenches, screwdrivers, old parts, old motorcycles, new parts, new motorcycles, sales literature, inner tubes, all scattered so thickly and clutteredly you can't even see the workbenches under them. I couldn't work in conditions like this but that's just because I'm not a photographic-mind mechanic. Bill can probably turn around and put his hand on any tool in this mess without having to think about where it is. I've seen mechanics like that. Drive you crazy to watch them, but they get the job done just as well and sometimes faster. Move one tool three inches to the left though, and he'll have to spend days looking for it.

Bill arrives with a grin about something. Sure, he's got some jets for my machine and knows right where they are. I'll have to wait a second though. He's got to close a deal out in back on some Harley parts. I go with him out in a shed in back and see he is selling a whole Harley machine in used parts, except for the frame, which the customer already has. He is selling them all for $125. Not a bad price at all.

Coming back I comment, "He'll know something about motorcycles before he gets *those* together."

Bill laughs. "And that's the best way to learn, too."

He has the jets and foot-peg rubber but no chain adjuster link. I get the rubber and jets installed, take the lump out of the idle and ride back to the hotel.

Sylvia and John and Chris are just coming down the stairs with their stuff as I arrive. Their faces indicate they're in the same good mood I'm in. We head down the main street, find a restaurant and order steaks for lunch.

"This is a great town," John says, "really *great*. Surprised there were any like this left. I was looking all over this morning. They've got stockmen's bars, high-top boots, silver-dollar belt buckles, Levis, Stetsons, the whole thing . . . and it's *real*. It isn't just Chamber of Commerce stuff. . . . In the bar down the block this morning they just started talking to me like I'd lived here all my life."

We order a round of beers. I see by a horseshoe sign on the wall we're into Olympia beer territory now, and order that.

"They must have thought I was off a ranch or something," John continues. "And this one old guy was talking away about how he wasn't going to give a thing to the goddam boys, and I really enjoyed that. The ranch was going to go to the girls, 'cause the goddam boys spend every cent they got down at Suzie's." John breaks up with laughter. "Sorry he ever raised 'em, and so on. I thought all that stuff disappeared thirty years ago, but it's still here."

The waitress comes with the steaks and we knife right into them. That work on the cycle has given me an appetite.

"Something else that ought to interest you," John says. "They were talking in the bar about Bozeman, where we're going. They said the governor of Montana had a list of fifty radical college professors at the college in Bozeman he was going to fire. Then he got killed in a plane crash."

"That was a long time ago," I answer. These steaks really *are* good.

"I didn't know they *had* a lot of radicals in this state."

"They've got *all* kinds of people in this state," I say. "But that was just right-wing politics."

John helps himself to some more salt. He says, "A Washington newspaper columnist came through and put it in his column yesterday, and that's why they were all talking about it. The president of the college confirmed it."

"Did they print the list?"

"I don't know. Did you know any of them?"

"If they had fifty names," I say, "mine must have been one."

They both look at me with some surprise. I don't know much about it, actually. It was *him,* of course, and with some feeling of falseness because of this I explain that a "radical" in Gallatin County, Montana, is a little different from a radical somewhere else.

"This was a college," I tell them, "where the wife of the president of the United States was actually banned because she was 'too controversial.' "

"Who?"

"Eleanor Roosevelt."

"Oh my God," John laughs, "that must have been *wild.*"

They want to hear more but it's hard to say anything. Then I remember one thing: "In a situation like that a *real* radical's actually got a perfect setup. He can do almost anything and get away with it because his opposition have already made asses out of themselves. They'll make him look good no matter what he says."

On the way out we pass a city park which I noticed last night, and which produced a memory concurrence. Just a vision of looking up into some trees. He had slept on that park bench one night on his way through to Bozeman. That's why I didn't recognize that forest yesterday. He'd come through at night, on his way to the college at Bozeman.

9

Now we follow the Yellowstone Valley right across Montana. It changes from Western sagebrush to Midwestern cornfields and back again, depending on whether it's under irrigation from the river. Sometimes we cross over bluffs that take us out of the irrigated area, but usually we stay close to the river. We pass by a marker saying something about Lewis and Clark. One of them came up this way on a side excursion from the Northwest Passage.

Nice sound. Fits the Chautauqua. We're really on a kind of Northwest Passage too. We pass through more fields and desert and the day wears on.

I want to pursue further now that same ghost that Phaedrus pursued—rationality itself, that dull, complex, classical ghost of underlying form.

This morning I talked about hierarchies of thought—the system. Now I want to talk about methods of finding one's way through these hierarchies—logic.

Two kinds of logic are used, inductive and deductive. Inductive inferences start with observations of the machine and arrive at general conclusions. For example, if the cycle goes over a bump and the engine misfires, and then goes over another bump and the engine misfires, and then goes over another bump and the engine misfires, and then goes over a long smooth stretch of road and there is no misfiring, and then goes over a fourth bump and the engine misfires again, one can logically conclude that the misfiring is caused by the bumps. That is induction: reasoning from particular experiences to general truths.

Deductive inferences do the reverse. They start with general knowledge and predict a specific observation. For example, if, from reading the hierarchy of facts about the machine, the mechanic knows the horn of the cycle is powered exclusively by electricity from the battery, then he can logically infer that if the battery is dead the horn will not work. That is deduction.

Solution of problems too complicated for common sense to solve is achieved by long strings of mixed inductive and deductive inferences that weave back and forth between the observed machine and the mental hierarchy of the machine found in the manuals. The correct program for this interweaving is formalized as scientific method.

Actually I've never seen a cycle-maintenance problem complex enough really to require full-scale formal scientific method. Repair problems are not that hard. When I think of formal scientific method an image sometimes comes to mind of an enormous juggernaut, a huge bulldozer—slow, tedious, lumbering, laborious, but invincible. It takes twice as long, five times as long, maybe a dozen times as long as informal mechanic's techniques, but you know in the end you're going to *get it.* There's no fault isolation problem in motorcycle maintenance that can stand up to it. When you've hit a really tough one, tried everything, racked your brain and nothing works, and you know that this time Nature has really decided to be difficult, you say, "Okay, Nature, that's the end of the *nice* guy," and you crank up the formal scientific method.

For this you keep a lab notebook. Everything gets written down, formally, so that you know at all times where you are, where you've been, where you're going and where you want to get. In scientific work and electronics technology this is necessary because otherwise the problems get so complex you get lost in them and confused and forget what you know and what you don't know and have to give up. In cycle maintenance things are not that involved, but when confusion starts it's a good idea to hold it down by making everything formal and exact. Sometimes just the act of writing down the problems straightens out your head as to what they really are.

The logical statements entered into the notebook are broken down into six categories: (1) statement of the problem, (2) hypotheses as to the cause of the problem, (3) experiments designed to test each hypothesis, (4) predicted results of the experiments, (5) observed results of the experiments and (6) conclusions from the results of the experiments. This is not different from the formal arrangement of many college and high-school lab notebooks but the purpose here is no longer just busywork. The purpose now is precise guidance of thoughts that will fail if they are not accurate.

The real purpose of scientific method is to make sure Nature hasn't misled you into thinking you know something you don't actually know. There's not a mechanic or scientist or technician alive who hasn't suffered from that one so much that he's not instinctively on guard. That's the main reason why so much scientific and mechanical information sounds so dull and so cautious. If you get careless or go romanticizing scientific information, giving it a flourish here and there, Nature will soon make a complete fool out of you. It does it often enough anyway even when you don't give it opportunities. One must be extremely careful and rigidly logical when dealing with Nature: one logical slip and an entire scientific edifice comes tumbling down. One false deduction about the machine and you can get hung up indefinitely.

In Part One of formal scientific method, which is the statement of the problem, the main skill is in stating absolutely no more than you are positive you know. It is much better to enter a statement "Solve Problem: Why doesn't cycle work?" which sounds dumb but is correct, than it is to enter a statement "Solve Problem: What is wrong with the electrical system?" when you don't absolutely *know* the trouble is *in* the electrical system. What you should state is "Solve Problem: What is wrong with cycle?" and *then* state as the first entry of Part Two: "Hypothesis Number One: The trouble is in the electrical system." You think of as many hypotheses as you can, then you design experiments to test them to see which are true and which are false.

This careful approach to the beginning questions keeps you from taking a major wrong turn which might cause you weeks of extra work or can even hang you up completely. Scientific questions often have a surface appearance of dumbness for this reason. They are asked in order to prevent dumb mistakes later on.

Part Three, that part of formal scientific method called experimentation, is sometimes thought of by romantics as all of science itself because that's the only part with much visual surface. They see lots of test tubes and bizarre equipment and people running around making discoveries. They do not see the experiment as part of a larger intellectual process and so they often confuse experiments with demonstrations, which look the same. A man conducting a gee-whiz science show with fifty thousand dollars' worth of Frankenstein equipment is not doing anything scientific if he knows beforehand what the results of his efforts are going to be. A motorcycle mechanic, on the other hand, who honks the horn to see if the battery works is informally conducting a true scientific experiment. He is testing a hypothesis by putting the question to nature. The TV scientist who mutters sadly, "The experiment is a failure; we have failed to achieve what we had hoped for," is suffering mainly from a bad scriptwriter. An experiment is never a failure solely because it fails to achieve predicted results. An experiment is a failure only when it also fails adequately to test the hypothesis in question, when the data it produces don't prove anything one way or another.

Skill at this point consists of using experiments that test only the hypothesis in question, nothing less, nothing more. If the horn honks, and the mechanic concludes that the whole electrical system is working, he is in deep trouble. He has reached an illogical conclusion. The honking horn only tells him that the battery and horn are working. To design an experiment properly he has to think very rigidly in terms of what directly causes what. This you know from the hierarchy. The horn doesn't make the cycle go. Neither does the battery, except in a very indirect way. The point at which the electrical system *directly* causes the engine to fire is at the spark plugs, and if you don't test here, at the output of the electrical system, you will never really know whether the failure is electrical or not.

To test properly the mechanic removes the plug and lays it against the engine so that the base around the plug is electrically grounded, kicks the starter lever and watches the spark-plug gap for a blue spark. If there isn't any he can conclude one of two things: (a) there is an electrical failure or (b) his experiment is sloppy. If he is experienced he will try it a few more times, checking connections, trying every way he can think of to get that plug to fire. Then, if he can't get it to fire, he finally concludes that *a* is correct, there's an electrical failure, and the experiment is over. He has proved that his hypothesis is correct.

In the final category, conclusions, skill comes in stating no more than the experiment has proved. It hasn't proved that when he fixes the electrical system the motorcycle will start. There may be other things wrong. But he does know that the motorcycle isn't going to run until the electrical system is working and he sets up the next formal question: "Solve problem: what is wrong with the electrical system?"

He then sets up hypotheses for these and tests them. By asking the right question and choosing the right tests and drawing the right conclusions the mechanic works his way down the echelons of the motorcycle hierarchy until he has found the exact specific cause or causes of the engine failure, and then he changes them so that they no longer cause the failure.

An untrained observer will see only physical labor and often get the idea that physical labor is mainly what the mechanic does. Actually the physical labor is the smallest and easiest part of what the mechanic does. By far the greatest part of his work is careful observation and precise thinking. That is why mechanics sometimes seem so taciturn and withdrawn when performing tests. They don't like it when you talk to them because they are concentrating on mental images, hierarchies, and not really looking at you or the physical motorcycle at all. They are using the experiment as part of a program to expand their hierarchy of knowledge of the faulty motorcycle and compare it to the correct hierarchy in their mind. They are looking at underlying form.

A car with a trailer coming our way is passing and having trouble getting back into his lane. I flash my headlight to make sure he sees us. He sees us but he can't get back in. The shoulder is narrow and bumpy. It'll spill us if we take it. I'm braking, honking, flashing. Christ Almighty, he panics and heads for our shoulder! I hold steady to the edge of the road. Here he COMES! At the last moment he goes back and misses us by inches.

A cardboard carton flaps and rolls on the road ahead of us, and we watch it for a long time before we come to it. Fallen off somebody's truck evidently.

Now the shakes come. If we'd been in a car that would've been a head-on. Or a roll in the ditch.

We pull off into a little town that could be in the middle of Iowa. The corn is growing high all around and the smell of fertilizer is heavy in the air. We retreat from the parked cycles into an enormous, high-ceilinged old place. To go with the beer this time I order every kind of snack they've got, and we have a late lunch on peanuts, popcorn, pretzels, potato chips, dried anchovies, dried smoked fish of some other kind with a lot of fine little bones in it, Slim Jims, Long Johns, pepperoni, Fritos, Beer Nuts, ham-sausage spread, fried pork rind and some sesame crackers with an extra taste I'm unable to identify.

Sylvia says, "I'm still feeling weak."

She somehow thought that cardboard box was our motorcycle rolling over and over again on the highway.

10

Outside in the valley again the sky is still limited by the bluffs on either side of the river, but they are closer together and closer to us than they were this morning. The valley is narrowing as we move toward the river's source.

We're also at a kind of beginning point in the things I'm discussing at which one can at last start to talk about Phaedrus' break from the mainstream of rational thought in pursuit of the ghost of rationality itself.

There was a passage he had read and repeated to himself so many times it survives intact. It begins:

> In the temple of science are many mansions . . . and various indeed are they that dwell therein and the motives that have led them there.
>
> Many take to science out of a joyful sense of superior intellectual power; science is their own special sport to which they look for vivid experience and the satisfaction of ambition; many others are to be found in the temple

who have offered the products of their brains on this altar for purely utilitarian purposes. Were an angel of the Lord to come and drive all the people belonging to these two categories out of the temple, it would be noticeably emptier but there would still be some men of both present and past times left inside. . . . If the types we have just expelled were the only types there were, the temple would never have existed any more than one can have a wood consisting of nothing but creepers . . . those who have found favor with the angel . . . are somewhat odd, uncommunicative, solitary fellows, really less like each other than the hosts of the rejected.

What has brought them to the temple . . . no single answer will cover . . . escape from everyday life, with its painful crudity and hopeless dreariness, from the fetters of one's own shifting desires. A finely tempered nature longs to escape from his noisy cramped surroundings into the silence of the high mountains where the eye ranges freely through the still pure air and fondly traces out the restful contours apparently built for eternity.

The passage is from a 1918 speech by a young German scientist named Albert Einstein.

Phaedrus had finished his first year of University science at the age of fifteen. His field was already biochemistry, and he intended to specialize at the interface between the organic and inorganic worlds now known as molecular biology. He didn't think of this as a career for his own personal advancement. He was very young and it was a kind of noble idealistic goal.

The state of mind which enables a man to do work of this kind is akin to that of the religious worshipper or lover. The daily effort comes from no deliberate intention or program, but straight from the heart.

If Phaedrus had entered science for ambitious or utilitarian purposes it might never have occurred to him to ask questions about the nature of a scientific hypothesis as an entity in itself. But he did ask them, and was unsatisfied with the answers.

The formation of hypotheses is the most mysterious of all the categories of scientific method. Where they come from, no one knows. A person is sitting somewhere, minding his own business, and suddenly— flash!—he understands something he didn't understand before. Until it's tested the hypothesis isn't truth. For the tests aren't its source. Its source is somewhere else. Einstein had said:

> Man tries to make for himself in the fashion that suits him best a simplified and intelligible picture of the world. He then tries to some extent to substitute this cosmos of his for the world of experience, and thus to overcome it. . . . He makes this cosmos and its construction the pivot of his emotional life in order to find in this way the peace and serenity which he cannot find in the narrow whirlpool of personal experience. . . . The supreme task . . . is to arrive at those universal elementary laws from which the cosmos can be built up by pure deduction. There is no logical path to these laws; only intuition, resting on sympathetic understanding of experience, can reach them. . . .

Intuition? Sympathy? Strange words for the origin of scientific knowledge.

A lesser scientist than Einstein might have said, "But scientific knowledge comes from *nature*. *Nature* provides the hypotheses." But Einstein understood that nature does not. Nature provides only experimental data.

A lesser mind might then have said, "Well then, *man* provides the hypotheses." But Einstein denied this too. "Nobody," he said, "who has really gone into the matter will deny that in practice the world of phenomena uniquely determines the theoretical system, in spite of the fact that there is no theoretical bridge between phenomena and their theoretical principles."

Phaedrus' break occurred when, as a result of laboratory experience, he became interested in hypotheses as entities in themselves. He had noticed again and again in his lab work that what might seem to be the hardest part of scientific work, thinking up the hypotheses, was invariably the easiest. The act of formally writing everything down precisely and clearly seemed to suggest them. As he was testing hypothesis number one by experimental method a flood of other hypotheses would come to mind, and as he was testing these, some more came to mind, and as he was testing these, still more came to mind until it became painfully evident that as he continued testing hypotheses and eliminating them or confirming them their number did not decrease. It actually *increased* as he went along.

At first he found it amusing. He coined a law intended to have the humor of a Parkinson's law that "The number of rational hypotheses that can explain any given phenomenon is infinite." It pleased him never to run out of hypotheses. Even when his experimental work seemed dead-end in every conceivable way, he knew that if he just sat down and muddled about it long enough, sure enough, another hypothesis would come along. And it always did. It was only months after he had coined the law that he began to have some doubts about the humor or benefits of it.

If true, that law is not a minor flaw in scientific reasoning. The law is completely nihilistic. It is a catastrophic logical disproof of the general validity of all scientific method!

If the purpose of scientific method is to select from among a multitude of hypotheses, and if the number of hypotheses grows faster than experimental method can handle, then it is clear that all hypotheses can never be tested. If all hypotheses cannot be tested, then the results of any experiment are inconclusive and the entire scientific method falls short of its goal of establishing proven knowledge.

About this Einstein had said, "Evolution has shown that at any given moment out of all conceivable constructions a single one has always proved itself absolutely superior to the rest," and let it go at that. But to Phaedrus that was an incredibly weak answer. The phrase "at any given moment" really shook him. Did Einstein really mean to state that truth was a function of time? To state *that* would annihilate the most basic presumption of all science!

But there it was, the whole history of science, a clear story of continuously new and changing explanations of old facts. The time spans of permanence seemed completely random, he could see no order in them. Some scientific truths seemed to last for centuries, others for less than a year. Scientific truth was not dogma, good for eternity, but a temporal quantitative entity that could be studied like anything else.

He studied scientific truths, then became upset even more by the apparent cause of their temporal condition. It looked as though the time spans of scientific truths are an inverse function of the intensity of scientific effort. Thus the scientific truths of the twentieth century seem to have a much shorter life-span than those of the last century because scientific activity is now much greater. If, in the next century, scientific activity increases tenfold, then the life expectancy of any scientific truth can be expected to drop to perhaps one-tenth as long as now. What shortens the life-span of the existing truth is the volume of hypotheses offered to replace it; the more the hypotheses, the shorter the time span of the truth. And what seems to be causing the number of hypotheses to grow in recent decades seems to be nothing other than scientific method itself. The more you look, the more you see. Instead of selecting one truth from a multitude you are *increasing the multitude*. What this means logically is that as you try to move toward unchanging truth through the application of scientific method, you actually do not move toward it at all. You move *away* from it! It is your application of scientific method that is causing it to change!

What Phaedrus observed on a personal level was a phenomenon, profoundly characteristic of the history of science, which has been swept under the carpet for years. The predicted results of scientific inquiry and the actual results of scientific inquiry are diametrically opposed here, and no one seems to pay too much attention to the fact. The purpose of scientific method is to select a single truth from among many hypothetical truths. That, more than anything else, is what science is all about. But historically science has done exactly the opposite. Through multiplication upon multiplication of facts, information, theories and hypotheses, it is science itself that is leading mankind from single absolute truths to multiple, indeterminate, relative ones. The major producer of the social chaos, the indeterminacy of thought and values that rational knowledge is supposed to eliminate, is none other than science itself. And what Phaedrus saw in the isolation of his own laboratory work years ago is now seen everywhere in the technological world today. Scientifically produced antiscience—chaos.

It's possible now to look back a little and see why it's important to talk about this person in relation to everything that's been said before concerning the division between classic and romantic realities and the irreconcilability of the two. Unlike the multitude of romantics who are disturbed about the chaotic changes science and technology force upon the human spirit, Phaedrus, with his scientifically trained classic mind, was able to do more than just wring his hands with dismay, or run away, or condemn the whole situation broadside without offering any solutions.

As I've said, he did in the end offer a number of solutions, but the problem was so deep and so formidable and complex that no one really understood the gravity of what he was resolving, and so failed to understand or misunderstood what he said.

The cause of our current social crises, he would have said, is a genetic defect within the nature of reason itself. And until this genetic defect is cleared, the crises will continue. Our current modes of rationality are not moving society forward into a better world. They are taking it further and further from that better world. Since the Renaissance these modes have worked. As long as the need for food, clothing and shelter is dominant they will continue to work. But now that for huge masses of people these needs no longer overwhelm everything else, the whole structure of reason, handed down to us from ancient times, is no longer adequate. It begins to be seen for what it really is—emotionally hollow, esthetically meaningless and spiritually empty. That, today, is where it is at, and will continue to be at for a long time to come.

I've a vision of an angry continuing social crisis that no one really understands the depth of, let alone has solutions to. I see people like John and Sylvia living lost and alienated from the whole rational structure of civilized life, looking for solutions outside that structure, but finding none that are really satisfactory for long. And then I've a vision of Phaedrus and his lone isolated abstractions in the laboratory—actually concerned with the same crisis but starting from another point, moving in the opposite direction—and what I'm trying to do here is put it all together. It's so big—that's why I seem to wander sometimes.

No one that Phaedrus talked to seemed really concerned about this phenomenon that so baffled him. They seemed to say, "We know scientific method is valid, so why ask about it?"

Phaedrus didn't understand this attitude, didn't know what to do about it, and because he wasn't a student of science for personal or utilitarian reasons, it just stopped him completely. It was as if he were contemplating that serene mountain landscape Einstein had described, and suddenly between the mountains had appeared a fissure, a gap of pure nothing. And slowly, and agonizingly, to explain this gap, he had to admit that the mountains, which had seemed built for eternity, might possibly be something else . . . perhaps just figments of his own imagination. It stopped him.

And so Phaedrus, who at the age of fifteen had finished his freshman year of science, was at the age of seventeen expelled from the University for failing grades. Immaturity and inattention to studies were given as official causes.

There was nothing anyone could have done about it; either to prevent it or correct it. The University couldn't have kept him on without abandoning standards completely.

In a stunned state Phaedrus began a long series of lateral drifts that led him into a far orbit of the mind, but he eventually returned along a route we are now following, to the doors of the University itself. Tomorrow I'll try to start on that route.

At Laurel, in sight of the mountains at last, we stop for the night. The evening breeze is cool now. It comes down off the snow. Although the sun must have disappeared behind the mountains an hour ago, there's still good light in the sky from behind the range.

Sylvia and John and Chris and I walk up the long main street in the gathering dusk and feel the presence of the mountains even though we talk about other things. I feel happy to be here, and still a little sad to be here too. Sometimes it's a little better to travel than to arrive.

11

I wake up wondering if I know we're near mountains because of memory or because of something in the air. We're in a beautiful old wooden room of a hotel. The sun is shining on the dark wood through the window shade, but even with the shade drawn I can sense that we're near mountains. There's mountain air in this room. It's cool and moist and almost fragrant. One deep breath makes me ready for the next one and then the next one and with each deep breath I feel a little readier until I jump out of bed and pull up the shade and let all that sunlight in—brilliant, cool, bright, sharp and clear.

An urge grows to go over and push Chris up and down to bounce him awake to see all this, but out of kindness, or respect maybe, he is allowed to sleep a while longer, and so with razor and soap I go to a common washroom at the other end of a long corridor of the same dark wood, floorboards creaking all the way. In the washroom the hot water is steaming and perking in the pipes, too hot at first for shaving, but fine after I mix it with cold water.

Through the window beyond the mirror I see there is a porch out in back, and when done go out and stand on it. It's at a level with the tops of the trees surrounding the hotel which seem to respond to this morning air the same way I do. The branches and leaves move with each light breeze as if it were expected, were what had been waited for all this time.

Chris is soon up and Sylvia comes out of her room saying she and John have already eaten breakfast and he is out walking somewhere, but she will go with Chris and me and walk down with us to breakfast.

We are in love with everything this morning and talk about good things all the way down a sunlit morning street to a restaurant. The eggs and hot cakes and coffee are from heaven. Sylvia and Chris talk intimately about his school and friends and personal things, while I listen and gaze through the large restaurant window at the storefront across the road. So different now from that lonely night in South Dakota. Beyond those buildings are mountains and snowfields.

Sylvia says John has talked to someone in town about another route to Bozeman, south through Yellowstone Park.

"South?" I say. "You mean Red Lodge?"

"I guess so."

A memory comes to me of snowfields in June. "That road goes way up above the timberline."

"Is that bad?" Sylvia asks.

"It'll be cold." In the middle of the snowfields in my mind appear the cycles and us riding on them. "But just tremendous."

We meet John again and it's settled. Soon, beyond a railroad underpass, we are on a twisting blacktop through fields toward the mountains up ahead. This is a road Phaedrus used all the time, and flashes of his memory coincide everywhere. The high, dark Absaroka Range looms directly ahead.

We are following a creek to its source. It contains water that was probably snow less than an hour ago. The stream and the road pass through green and stony fields each a little higher than before. Everything is so intense in this sunlight. Dark shadows, bright light. Dark blue sky. The sun is bright and hot when we're in it, but when we pass under trees along the road, it's suddenly cold.

We play tag with a little blue Porsche along the way, passing it with a beep and being passed by it with a beep and doing this several times through fields of dark aspen and bright greens of grass and mountain shrubs. All this is remembered.

He would use this route to get into the high country, then backpack in from the road for three or four or five days, then come back out for more food and head back in again, needing these mountains in an almost physiological way. The train of his abstractions became so long and so involved he had to have the surroundings of silence and space here to hold it straight. It was as though hours of constructions would have been shattered by the least distraction of other thought or other duty. It wasn't like other people's thinking, even then, before his insanity. It was at a level at which everything shifts and changes, at which institutional values and verities are gone and there is nothing but one's own spirit to keep one going. His early failure had re-

leased him from any felt obligation to think along institutional lines and his thoughts were already independent to a degree few people are familiar with. He
felt that institutions such as schools, churches, governments and political organizations of every sort all
tended to direct thought for ends other than truth, for
the perpetuation of their own functions, and for the
control of individuals in the service of these functions.
He came to see his early failure as a lucky break, an ac-
cidental escape from a trap that had been set for him,
and he was very trap-wary about institutional truths for
the remainder of his time. He didn't see these things
and think this way at first, however, only later on. I'm
getting way out of sequence here. This all came much
later.

At first the truths Phaedrus began to pursue were lateral truths; no longer the frontal truths of science,
those toward which the discipline pointed, but the kind
of truth you see laterally, out of the corner of your eye.
In a laboratory situation, when your whole procedure
goes haywire, when everything goes wrong or is indeterminate or is so screwed up by unexpected results
you can't make head or tail out of anything, you start
looking *laterally.* That's a word he later used to describe a growth of knowledge that doesn't move forward like an arrow in flight, but expands sideways, like
an arrow enlarging in flight, or like the archer, discovering that although he has hit the bull's-eye and won
the prize, his head is on a pillow and the sun is coming
in the window. Lateral knowledge is knowledge that's
from a wholly unexpected direction, from a direction
that's not even understood as a direction until the
knowledge forces itself upon one. Lateral truths point

to the falseness of axioms and postulates underlying one's existing system of getting at truth.

To all appearances he was just drifting. In actuality he was just drifting. Drifting is what one does when looking at lateral truth. He couldn't follow any known method of procedure to uncover its cause because it was these methods and procedures that were all screwed up in the first place. So he drifted. That was all he could do.

The drift took him into the Army, which sent him to Korea. From his memory there's a fragment, a picture of a wall, seen from a prow of a ship, shining radiantly, like a gate of heaven, across a misty harbor. He must have valued the fragment greatly and thought about it many times because although it doesn't fit anything else it is intense, so intense I've returned to it myself many times. It seems to symbolize something very important, a turning point.

His letters from Korea are radically different from his earlier writing, indicating this same turning point. They just explode with emotion. He writes page after page about tiny details of things he sees: marketplaces, shops with sliding glass doors, slate roofs, roads, thatched huts, everything. Sometimes full of wild enthusiasm, sometimes depressed, sometimes angry, sometimes even humorous, he is like someone or some creature that has found an exit from a cage he did not even know was around him, and is wildly roaming over the countryside visually devouring everything in sight.

Later he made friends with Korean laborers who spoke some English but wanted to learn more so that they could qualify as translators. He spent time with them after working hours and in return they took him on long weekend hikes through the hills to see their homes and friends and translate for him the way of life and thought of another culture.

He is sitting by a footpath on a beautiful windswept hillside overlooking the Yellow Sea. The rice in the terrace below the footpath is full-grown and brown. His friends look down at the sea with him seeing islands far out from shore. They eat a picnic lunch and talk to one another and to him and the subject is ideographs and their relation to the world. He comments on how amazing it is that everything in the universe can be described by the twenty-six written characters with which they have been working. His friends nod and smile and eat the food they've taken from tins and say no pleasantly.

He is confused by the nod yes and the answer no and so repeats the statement. Again comes the nod meaning yes and the answer no. That is the end of the fragment, but like the wall it's one he thinks about many times.

The final strong fragment from that part of the world is of a compartment of a troopship. He is on his way home. The compartment is empty and unused. He is alone on a bunk made of canvas laced to a steel frame, like a trampoline. There are five of these to a tier, tier after tier of them, completely filling the empty troop compartment.

This is the foremost compartment of the ship and the canvas in the adjoining frames rises and falls, accompanied by elevator feelings in his stomach. He contemplates these things and a deep booming on the steel plates all around him and realizes that except for these signs there is no indication whatsoever that this entire compartment is rising massively high up into the air and then plunging down, over and over again. He wonders if it is that which is making it difficult to concentrate on the book before him, but realizes that no, the book is just hard. It's a text on Oriental philosophy and it's the most difficult book he's ever read. He's glad to be alone and bored in this empty troop compartment, otherwise he'd never get through it.

The book states that there's a theoretic component of man's existence which is primarily Western (and this corresponded to Phaedrus' laboratory past) and an esthetic component of man's existence which is seen more strongly in the Orient (and this corresponded to Phaedrus' Korean past) and that these never seem to meet. These terms "theoretic" and "esthetic" correspond to what Phaedrus later called classic and romantic modes of reality and probably shaped these terms in his mind more than he ever knew. The difference is that the classic reality is *primarily* theoretic but has its own esthetics too. The romantic reality is *primarily* esthetic, but has its theory too. The theoretic and esthetic split is between components of a single world. The classic and romantic split is between two separate worlds. The philosophy book, which is called *The Meeting of East and West,* by F. S. C. Northrop, suggests that greater cognizance be made of the "undifferentiated aesthetic continuum" from which the theoretic arises.

Phaedrus didn't understand this, but after arriving in Seattle, and his discharge from the Army, he sat in his hotel room for two whole weeks, eating enormous Washington apples, and thinking, and eating more apples, and thinking some more, and then as a result of all these fragments, and thinking, returned to the University to study philosophy. His lateral drift was ended. He was actively in pursuit of something now.

A sudden cross-gust of cold air comes heavy with the smell of pines, and soon another and another, and as we approach Red Lodge I'm shivering.

At Red Lodge the road's almost joined to the base of the mountain. The dark ominous mass beyond dominates even the roofs of the buildings on either side of the main street. We park the cycles and unpack them to remove warm clothing. We walk past ski shops into a restaurant where we see on the walls huge photographs of the route we will take up. And up and up, over one of the highest paved roads in the world. I feel some anxiety about this, which I realize is irrational and try to get rid of by talking about the road to the others. There's no way to fall off. No danger to the motorcycle. Just a memory of places where you could throw a stone and it would drop thousands of feet before coming to rest and somehow associating that stone with the cycle and rider.

When coffee is finished we put on the heavy clothing, repack and have soon traveled to the first of many switchback turns across the face of the mountain.

The asphalt of the road is much wider and safer than it occurred in memory. On a cycle you have all sorts of extra room. John and Sylvia take the hairpin turns up ahead and then come back above us, facing us, and have smiles. Soon we take the turn and see their backs again. Then another turn for them and we meet them again, laughing. It's so hard when contemplated in advance, and so easy when you do it.

I talked about Phaedrus' lateral drift, which ended with entry into the discipline of philosophy. He saw philosophy as the highest echelon of the entire hierarchy of knowledge. Among philosophers this is so widely believed it's almost a platitude, but for him it's a revelation. He discovered that the science he'd once thought of as the whole world of knowledge is only a branch of philosophy, which is far broader and far more general. The questions he had asked about infinite hypotheses hadn't been of interest to science because they weren't scientific questions. Science cannot study scientific method without getting into a bootstrap problem that destroys the validity of its answers. The questions he'd asked were at a higher level than science goes. And so Phaedrus found in philosophy a natural continuation of the question that brought him to science in the first place. What does it all mean? What's the purpose of all this?

At a turnout on the road we stop, take some record photographs to show we have been here and then walk to a little path that takes us out to the edge of a cliff. A motorcycle on the road almost straight down beneath us could hardly be seen from up here. We bundle up more tightly against the cold and continue upward.

The broad-leafed trees are all gone. Only small pines are left. Many of these have twisted and stunted shapes.

Soon stunted pines disappear entirely and we're in

alpine meadows. There's not a tree anywhere, only grass everywhere filled with little pink and blue and white dots of intense color. Wildflowers, everywhere! These and grasses and mosses and lichens are all that can live here, now. We've reached the high country, above the timberline.

I look over my shoulder for one last view of the gorge. Like looking down at the bottom of the ocean. People spend their entire lives at those lower altitudes without any awareness that this high country exists.

The road turns inward, away from the gorge and into snowfields.

The engine backfires fiercely from lack of oxygen and threatens to stop but never does. Soon we are between banks of old snow, the way snow looks in early spring after a thaw. Little streams of water run everywhere into mossy mud, and then below this into week-old grass and then small wildflowers, the tiny pink and blue and yellow and white ones which seem to pop out, sun-brilliant, from black shadows. Everywhere it's like this! Little pins of colored light shoot forth to me from a background of somber dark green and black. Dark sky now and cold. Except where the sun hits. On the sun side my arm and leg and jacket are hot, but the dark side, in deep shadows now, is very cold.

The snowfields become heavy and show steep banks where snowplows have been. The banks become four feet high, then six feet, then twelve feet high. We move through twin walls, almost a tunnel of snow. Then the tunnel opens onto dark sky again and when we emerge we see we're at the summit.

Beyond is another country. Mountain lakes and pines and snowfields are below. Above and beyond them as far as we can see are farther mountain ranges covered with snow. The high country.

We stop and park at a turnoff where a number of tourists take pictures and look around at the view and at one another. At the back of his cycle John removes his camera from the saddlebag. From my own machine I remove the tool kit and spread it out on the seat, then take the screwdriver, start the engine and with the screwdriver adjust the carburetors until the idling sound changes from a really bad loping to just slightly bad. I'm surprised at how all the way up it backfired and sputtered and kicked and gave every indication it was going to quit but never did. I didn't adjust them, out of curiosity to see what eleven thousand feet of altitude would do. Now I'm leaving them rich and sounding just bad because we'll be going down some now toward Yellowstone Park and if they aren't slightly rich now they'll get too lean later on, which is dangerous because it overheats the engine.

The backfiring is still fairly heavy on the way down from the summit with the engine dragging in second gear, but then the noise diminishes as we reach lower altitudes. The forests return. We move among rocks and lakes and trees now, taking beautiful turns and curves of the road.

I want to talk about another kind of high country now in the world of thought, which in some ways, for me at least, seems to parallel or produce feelings similar to this, and call it the high country of the mind.

If all of human knowledge, everything that's known, is believed to be an enormous hierarchic structure, then the high country of the mind is found at the uppermost reaches of this structure in the most general, the most abstract considerations of all.

Few people travel here. There's no real profit to be made from wandering through it, yet like this high country of the material world all around us, it has its own austere beauty that to some people make the hardships of traveling through it seem worthwhile.

In the high country of the mind one has to become adjusted to the thinner air of uncertainty, and to the enormous magnitude of questions asked, and to the answers proposed to these questions. The sweep goes on and on and on so obviously much further than the mind can grasp one hesitates even to go near for fear of getting lost in them and never finding one's way out.

What is the truth and how do you know it when you have it? . . . How do we really *know* anything? Is there an "I," a "soul," which knows, or is this soul merely cells coordinating senses? . . . Is reality basically changing, or is it fixed and permanent? . . . When it's said that something *means* something, what's meant by that?

Many trails through these high ranges have been made and forgotten since the beginning of time, and although the answers brought back from these trails have claimed permanence and universality for themselves, civilizations have varied in the trails they have chosen and we have many different answers to the same question, all of which can be thought of as true within their own context. Even within a single civilization old trails are constantly closed and new ones opened up.

It's sometimes argued that there's no real progress; that a civilization that kills multitudes in mass warfare, that pollutes the land and oceans with ever larger quantities of debris, that destroys the dignity of individuals by subjecting them to a forced mechanized existence can hardly be called an advance over the simpler hunting and gathering and agricultural existence of prehistoric times. But this argument, though romantically appealing, doesn't hold up. The primitive tribes permitted far less individual freedom than does modern society. Ancient wars were committed with far less moral justification than modern ones. A technology that produces debris can find, and is finding, ways of disposing of it without ecological upset. And the schoolbook pictures of primitive man sometimes omit some of the detractions of his primitive life—the pain, the disease, famine, the hard labor needed just to stay alive. From that agony of bare existence to modern life can be soberly described only as upward progress, and the sole agent for this progress is quite clearly reason itself.

One can see how both the informal and formal processes of hypothesis, experiment, conclusion, century after century, repeated with new material, have built up the hierarchies of thought which have eliminated most of the enemies of primitive man. To some extent the romantic condemnation of rationality stems from the very effectiveness of rationality in uplifting men from primitive conditions. It's such a powerful, all-dominating agent of civilized man it's all but shut out everything else and now dominates man himself. That's the source of the complaint.

Phaedrus wandered through this high country, aimlessly at first, following every path, every trail where someone had been before, seeing occasionally with small hindsights that he was apparently making some progress, but seeing nothing ahead of him that told him which way to go.

Through the mountainous questions of reality and knowledge had passed great figures of civilization, some of whom, like Socrates and Aristotle and Newton and Einstein, were known to almost everyone, but most of whom were far more obscure. Names he had never heard of before. And he became fascinated with their thought and their whole way of thinking. He followed their trails carefully until they seemed to grow cold, then dropped them. His work was just barely passing by academic standards at this time, but this wasn't because he wasn't working or thinking. He was thinking too hard, and the harder you think in this high country of the mind the slower you go. Phaedrus read in a scientific way rather than a literary way, testing each sentence as he went along, noting doubts and questions to be resolved later, and I'm fortunate in having a whole trunkful of volumes of these notations.

What is most astonishing about them is that almost everything he said years later is contained in them. It's frustrating to see how completely unaware he is at the time of the significance of what he is saying. It's like seeing someone handling, one by one, all the pieces of a jigsaw puzzle whose solution you know, and you want to tell him, "Look, this fits *here,* and this fits *here,*" but you can't tell him. And so he wanders blindly along one trail after another gathering one piece after another and wondering what to do with them, and you grit your teeth when he goes off on a false trail and are relieved when he comes back again, even though he is discouraged himself. "Don't worry," you want to tell him. "Keep going!"

But he's such an abominable scholar it must be through the kindness of his instructors that he passes at all. He prejudges every philosopher he studies. He always intrudes and imposes his own views upon the material he is studying. He is never fair. He's always partial. He wants each philosopher to go a certain way and becomes infuriated when he does not.

A fragment of memory is preserved of him sitting in a room at three and four in the morning with Immanuel Kant's famous *Critique of Pure Reason*, studying it as a chess player studies the openings of the tournament masters, trying to test the line of development against his own judgment and skill, looking for contradictions and incongruities.

Phaedrus is a bizarre person when contrasted to the twentieth-century Midwestern Americans who surround him, but when he is seen studying Kant he is less strange. For this eighteenth-century German philosopher he feels a respect that rises not out of agreement but out of appreciation for Kant's formidable logical fortification of his position. Kant is always superbly methodical, persistent, regular and meticulous as he scales that great snowy mountain of thought concerning what is in the mind and what is outside the mind. It is, for modern climbers, one of the highest peaks of all, and I want now to magnify this picture of Kant and show a little about how he thought and how Phaedrus thought about him in order to give a clearer picture of what the high country of the mind is like and also to prepare the way for an understanding of Phaedrus' thoughts.

Phaedrus' resolution of the entire problem of classic and romantic understanding occurred at first in this high country of the mind, and unless one understands the relation of this country to the rest of existence, the meaning and the importance of lower levels of what he said here will be underestimated or misunderstood.

To follow Kant one must also understand something about the Scottish philosopher David Hume. Hume had previously submitted that if one follows the strictest rules of logical induction and deduction from experience to determine the true nature of the world, one must arrive at certain conclusions. His reasoning followed lines that would result from answers to this question: Suppose a child is born devoid of all senses; he has no sight, no hearing, no touch, no smell, no taste—nothing. There's no way whatsoever for him to receive any sensations from the outside world. And suppose this child is fed intravenously and otherwise attended to and kept alive for eighteen years in this state of existence. The question is then asked: Does this eighteen-year-old person have a thought in his head? If so, where does it come from? How does he get it?

Hume would have answered that the eighteen-year-old had no thoughts whatsoever, and in giving this answer would have defined himself as an *empiricist,* one who believes all knowledge is derived exclusively from the senses. The scientific method of experimentation is carefully controlled empiricism. Common sense today is empiricism, since an overwhelming majority would agree with Hume, even though in other cultures and other times a majority might have differed.

The first problem of empiricism, if empiricism is believed, concerns the nature of "substance." If all our knowledge comes from sensory data, what exactly is this substance which is supposed to give off the sensory data itself? If you try to imagine what this substance is, apart from what is sensed, you'll find yourself thinking about nothing whatsoever.

Since all knowledge comes from sensory impressions and since there's no sensory impression of substance itself, it follows logically that there is no knowledge of substance. It's just something we imagine. It's entirely within our own minds. The idea that there's something out there giving off the properties we receive is just another of those common-sense notions similar to the common-sense notion children have that the earth is flat and parallel lines never meet.

Secondly, if one starts with the premise that all our knowledge comes to us through our senses, one must ask, From what sense data is our knowledge of causation received? In other words, what is the scientific empirical basis of causation itself?

Hume's answer is "None." There's *no* evidence for causation in our sensations. Like substance, it's just something we imagine when one thing repeatedly follows another. It has no real existence in the world we observe. If one accepts the premise that all knowledge comes to us through our senses, Hume says, then one must logically conclude that both "Nature" and "Nature's laws" are creations of our own imagination.

This idea that the entire world is within one's own mind could be dismissed as absurd if Hume had just thrown it out for speculation. But he was making it an airtight case.

To throw out Hume's conclusions was necessary, but unfortunately he had arrived at them in such a way that it was seemingly impossible to throw them out without abandoning empirical reason itself and retiring into some medieval predecessor of empirical reason. This Kant would not do. Thus it was Hume, Kant said, who "aroused me from my dogmatic slumbers" and caused him to write what is now regarded as one of the greatest philosophical treatises ever written, the *Critique of Pure Reason,* often the subject of an entire University course.

Kant is trying to save scientific empiricism from the consequences of its own self-devouring logic. He starts out at first along the path that Hume has set before him. "That all our knowledge begins with experience there can be no doubt," he says, but he soon departs from the path by denying that all components of knowledge come from the senses at the moment the sense data are received. "But though all knowledge begins *with* experience it doesn't follow that it arises *out* of experience."

This seems, at first, as though he is picking nits, but he isn't. As a result of this difference, Kant skirts right around the abyss of solipsism that Hume's path leads to and proceeds on an entirely new and different path of his own.

Kant says there are aspects of reality which are not supplied immediately by the senses. These he calls *a priori.*

An example of *a priori* knowledge is "time." You don't see time. Neither do you hear it, smell it, taste it or touch it. It isn't present in the sense data as they are received. Time is what Kant calls an "intuition," which the mind must supply as it receives the sense data.

The same is true of space. Unless we *apply* the concepts of space and time to the impressions we receive, the world is unintelligible, just a kaleidoscopic jumble of colors and patterns and noises and smells and pain and tastes without meaning. We sense objects in a certain way because of our application of *a priori* intuitions such as space and time, but we do not create these objects out of our imagination, as pure philosophical idealists would maintain. The forms of space and time are applied to data as they are received from the object producing them. The *a priori* concepts have their origins in human nature so that they're neither caused by the sensed object nor bring it into being, but provide a kind of *screening* function for what sense data we will accept. When our eyes blink, for example, our sense data tell us that the world has disappeared. But this is screened out and never gets to our consciousness because we have in our minds an *a priori* concept that the world has continuity. What we think of as reality is a continuous synthesis of elements from a fixed hierarchy of *a priori* concepts and the ever changing data of the senses.

Now stop and apply some of the concepts Kant has put forth to this strange machine, this creation that's been bearing us along through time and space. See our relation to it now, as Kant reveals it to us.

Hume has been saying, in effect, that everything I know about this motorcycle comes to me through my senses. It has to be. There's no other way. If I say it's made of metal and other substances, he asks, What's metal? If I answer that metal's hard and shiny and cold to the touch and deforms without breaking under blows from a harder material, Hume says those are all sights and sounds and touch. There's no substance. Tell me what metal is *apart* from these sensations. Then, of course, I'm stuck.

But if there's no substance, what can we say about the sense data we receive? If I hold my head to the left and look down at the handle grips and front wheel and map carrier and gas tank I get one pattern of sense data. If I move my head to the right I get another slightly different pattern of sense data. The two views are different. The angles of the planes and curves of the metal are different. The sunlight strikes them differently. If there's no logical basis for substance then there's no logical basis for concluding that what's produced these two views is the same motorcycle.

Now we've a real intellectual impasse. Our reason, which is supposed to make things more intelligible, seems to be making them less intelligible, and when reason thus defeats its own purpose something has to be changed in the structure of our reason itself.

Kant comes to our rescue. He says that the fact that there's no way of immediately sensing a "motorcycle," as distinguished from the colors and shapes a motorcycle produces, is no proof at all that there's no motorcycle there. We have in our minds an *a priori* motorcycle which has continuity in time and space and is capable of changing appearance as one moves one's head and is therefore not contradicted by the sense data one is receiving.

Hume's motorcycle, the one that makes no sense at all, will occur if our previous hypothetical bed patient, the one who has no senses at all, is suddenly, for one second only, exposed to the sense data of a motorcycle, then deprived of his senses again. Now, I think, in his mind he would have a Hume motorcycle, which provides him with no evidence whatsoever for such concepts as causation.

But, as Kant says, we are not that person. We have in our minds a very real *a priori* motorcycle whose existence we have no reason to doubt, whose reality can be confirmed anytime.

This *a priori* motorcycle has been built up in our minds over many years from enormous amounts of sense data and it is constantly changing as new sense data come in. Some of the changes in this specific *a priori* motorcycle I'm riding are very quick and transitory, such as its relationship to the road. This I'm monitoring and correcting all the time as we take these curves and bends in the road. As soon as the information's of no more value I forget it because there's more coming in that must be monitored. Other changes in this *a priori* are slower: Disappearance of gasoline from the tank. Disappearance of rubber from the tires. Loosening of bolts and nuts. Change of gap between brake shoes and drums. Other aspects of the motorcycle change so slowly they seem permanent—the paint job, the wheel bearings, the control cables—yet these are constantly changing too. Finally, if one thinks in terms of really large amounts of time even the frame is changing slightly from the road shocks and thermal changes and forces of internal fatigue common to all metals.

It's quite a machine, this *a priori* motorcycle. If you stop to think about it long enough you'll see that it's the main thing. The sense data confirm it but the sense data aren't *it*. The motorcycle that I believe in an *a priori* way to be outside of myself is like the money I believe I have in the bank. If I were to go down to the bank and ask to see my money they would look at me a little peculiarly. They don't have "my money" in any little drawer that they can pull open to show me. "My money" is nothing but some east-west and north-south magnetic domains in some iron oxide resting on a roll of tape in a computer storage bin. But I'm satisfied with this because I've faith that if I need the things that money enables, the bank will provide the means, through their checking system, of getting it. Similarly, even though my sense data have never brought up anything that could be called "substance" I'm satisfied that there's a capability within the sense data of achieving the things that substance is supposed to do, and that the sense data will continue to match the *a priori* motorcycle of my mind. I say for the sake of convenience that I've money in the bank and say for the sake of convenience that substances compose the cycle I'm riding on. The bulk of Kant's *Critique of Pure Reason* is concerned with how this *a priori* knowledge is acquired and how it is employed.

Kant called his thesis that our *a priori* thoughts are independent of sense data and screen what we see a "Copernican revolution." By this he referred to Copernicus' statement that the earth moves around the sun. Nothing changed as a result of this revolution, and yet everything changed. Or, to put it in Kantian terms, the objective world producing our sense data did not change, but our *a priori* concept of it was turned inside out. The effect was overwhelming. It was the acceptance of the Copernican revolution that distinguishes modern man from his medieval predecessors.

"What Copernicus did was take the existing *a priori* concept of the world, the notion that it was flat and fixed in space, and pose an alternative *a priori* concept of the world, that it's spherical and moves around the sun; and showed that *both* of the *a priori* concepts fitted the existing sensory data.

Kant felt he had done the same thing in metaphysics. If you presume that the *a priori* concepts in our heads are independent of what we see and actually screen what we see, this means that you are taking the old Aristotelian concept of scientific man as a passive observer, a "blank tablet," and truly turning this concept inside out. Kant and his millions of followers have maintained that as a result of this inversion you get a much more satisfying understanding of how we know things.

I've gone into this example in some detail, partly to show some of the high country in close perspective, but more to prepare for what Phaedrus did later. He too performed a Copernican inversion and as a result of this inversion produced a resolution of the separate worlds of classical and romantic understanding. And it seems to me that as a result it is possible to again get a much more satisfying understanding of what the world is all about.

Kant's metaphysics thrilled Phaedrus at first, but later it dragged and he didn't know exactly why. He thought about it and decided that maybe it was the Oriental experience. He had had the feeling of escape from a prison of intellect, and now this was just more of the prison again. He read Kant's esthetics with disappointment and then anger. The ideas expressed about the "beautiful" were themselves ugly to him, and the ugliness was so deep and pervasive he hadn't a clue as to where to begin to attack it or try to get around it. It seemed woven right into the whole fabric of Kant's world so deeply there was no escape from it. It wasn't just eighteenth-century ugliness or "technical" ugliness. All of the philosophers he was reading showed it. The whole university he was attending smelled of the same ugliness. It was everywhere, in the classroom, in the textbooks. It was in himself and he didn't know how or why. It was reason itself that was ugly and there seemed no way to get free.

12

❧

At Cooke City John and Sylvia look and sound happier than I have seen them in years, and we whack into our hot beef sandwiches with great whacks. I'm happy to hear and see all their high-country exuberance but don't comment much, just keep eating.

Outside the picture window across the road are huge pines. Many cars pass beneath them on their way to the park. We're a long way down from the timberline now. Warmer here but covered over with an occasional low cloud ready to drop rain.

I suppose if I were a novelist rather than a Chautauqua orator I'd try to "develop the characters" of John and Sylvia and Chris with action-packed scenes that would also reveal "inner meanings" of Zen and maybe Art and maybe even Motorcycle Maintenance. That would be quite a novel, but for some reason I don't feel quite up to it. They're friends, not characters, and as Sylvia herself once said, "I don't like being an object!" So a lot of things we know about one another I'm simply not going into. Nothing bad, but not really relevant to the Chautauqua. That's the way it should be with friends.

At the same time I think you can understand from the Chautauqua why I must always seem so reserved and remote to them. Once in a while they ask questions that seem to call for a statement of what the hell I'm always thinking about, but if I were to babble what's really on my mind about, say, the *a priori* presumption of the continuity of a motorcycle from second to second and do this without benefit of the entire edifice of the Chautauqua, they'd just be startled and wonder what's wrong. I really *am* interested in this continuity and the way we talk and think about it and so tend to get removed from the usual lunchtime situation and this gives an appearance of remoteness. It's a problem.

It's a problem of our time. The range of human knowledge today is so great that we're all specialists and the distance between specializations has become so great that anyone who seeks to wander freely among them almost has to forego closeness with the people around him. The lunchtime here-and-now stuff is a specialty too.

Chris seems to understand my remoteness better than they do, perhaps because he's more used to it and his relationship to me is such that he has to be more concerned. In his face I sometimes see a look of worry, or at least anxiety, and wonder why, and then discover that I'm angry. If I hadn't seen his expression, I might not have known it. Other times he's running and jumping all over the place and I wonder why and discover that it's because I'm in a good mood. Now I see he's a little nervous and answering a question that John had evidently directed at me. It's about the people we'll be staying with tomorrow, the DeWeeses.

I'm not sure what the question was but add, "He's a painter. He teaches fine arts at the college there, an abstract impressionist."

They ask how I came to know him and I have to answer that I don't remember which is a little evasive. I don't remember *anything* about him except fragments. He and his wife were evidently friends of Phaedrus' friends, and he came to know them that way.

They wonder what an engineering writer like myself would have in common with an abstract painter and I have to say again that I don't know. I mentally file through the fragments for an answer but none comes.

Their personalities were certainly different. Whereas photographs of Phaedrus' face during this period show alienation and aggression—a member of his department had half-jokingly called it a "subversive" look—some photographs of DeWeese from the same period show a face that is quite passive, almost serene, except for a mild questioning expression.

In my memory is a movie about a World War I spy who studied the behavior of a captured German officer (who looked exactly like him) by means of a one-way mirror. He studied him for months until he could imitate every gesture and nuance of speech. Then he pretended to be the escaped officer in order to infiltrate the German Army command. I remember the tension and excitement as he faced his first test with the officer's old friends to learn if they would see through his

imposture. Now I've some of the same feeling about
DeWeese, who'll naturally presume I'm the person he
once knew.

Outside a light mist has made the motorcycles wet. I
take out the plastic bubble from the saddlebag and attach it to the helmet. We'll be entering Yellowstone
Park soon.

The road ahead is foggy. It seems like a cloud has
drifted into the valley, which isn't really a valley at all
but more of a mountain pass.

I don't know how well DeWeese knew him, and what
memories he'll expect me to share. I've gone through
this before with others and have usually been able to
gloss over awkward moments. The reward each time
has been an expansion of knowledge about Phaedrus
that has greatly aided further impersonation, and
which over the years has supplied the bulk of the information I've been presenting here.

From what fragments of memory I have, Phaedrus
had a high regard for DeWeese because he didn't understand him. For Phaedrus, failure to understand
something created tremendous interest and DeWeese's
attitudes were fascinating. They seemed all haywire.
Phaedrus would say something he thought was pretty
funny and DeWeese would look at him in a puzzled
way or else take him seriously. Other times Phaedrus
would say something that was very serious and of deep
concern, and DeWeese would break up laughing, as
though he had cracked the cleverest joke he had ever
heard.

For example, there is the fragment of memory about

a dining-room table whose edge veneer had come loose and which Phaedrus had reglued. He held the veneer in place while the glue set by wrapping a whole ball of string around the table, round and round and round.

DeWeese saw the string and wondered what that was all about.

"That's my latest sculpture," Phaedrus had said. "Don't you think it kind of builds?"

Instead of laughing, DeWeese looked at him with amazement, studied it for a long time and finally said, "Where did you learn all *this*?"

For a second Phaedrus thought he was continuing the joke, but he was serious.

Another time Phaedrus was upset about some failing students. Walking home with DeWeese under some trees he had commented on it and DeWeese had wondered why he took it so personally.

"I've wondered too," Phaedrus had said, and in a puzzled voice had added, "I think maybe it's because every teacher tends to grade up students who resemble him the most. If your own writing shows neat penmanship you regard that more important in a student than if it doesn't. If you use big words you're going to like students who write with big words."

"Sure. What's wrong with that?" DeWeese had said.

"Well, there's something whacky here," Phaedrus had said, "because the students I like the most, the ones I really feel a sense of identity with, are all *failing*!"

DeWeese had completely broken up with laughter at this and left Phaedrus feeling miffed. He had seen it as a kind of scientific phenomenon that might offer clues leading to new understanding, and DeWeese had just laughed.

At first he thought DeWeese was just laughing at his unintended insult to himself. But that didn't fit because DeWeese wasn't a derogatory kind of person at all. Later he saw it was a kind of supertruth laugh. The best students always *are* flunking. *Every* good teacher knows that. It was a kind of laughter that destroys tensions produced by impossible situations and Phaedrus could have used some of it because at this time he was taking things way too seriously.

These enigmatic responses of DeWeese gave Phaedrus the idea that DeWeese had access to a huge terrain of hidden understanding. DeWeese always seemed to be *concealing* something. He was *hiding* something from him, and Phaedrus couldn't figure out what it was.

Then comes a strong fragment, the day when he discovered DeWeese seemed to have the same puzzled feeling about *him.*

A light switch in DeWeese's studio didn't work and he asked Phaedrus if he knew what was wrong with it. He had a slightly embarrassed, slightly puzzled smile on his face, like the smile of an art patron talking to a painter. The patron is embarrassed to reveal how little he knows but is smiling with the expectation of learning more. Unlike the Sutherlands, who *hate* technology, DeWeese is so far removed from it he didn't feel it any particular menace. DeWeese was actually a technology *buff,* a *patron* of the technologies. He didn't understand them, but he knew what he liked, and he always enjoyed learning more.

He had the illusion the trouble was in the wire near

the bulb because immediately upon toggling the switch
the light went out. If the trouble had been in the switch,
he felt, there would have been a lapse of time before
the trouble showed up in the bulb. Phaedrus did not
argue with this, but went across the street to the hardware store, bought a switch and in a few minutes had
it installed. It worked immediately, of course, leaving
DeWeese puzzled and frustrated. "How did you know
the trouble was in the switch?" he asked.

"Because it worked intermittently when I jiggled the
switch."

"Well—couldn't it jiggle the wire?"

"No."

Phaedrus' cocksure attitude angered DeWeese and
he started to argue. "How do you *know* all that?" he
said.

"It's *obvious.*"

"Well then, why didn't I see it?"

"You have to have some familiarity."

"Then it's *not* obvious, is it?"

DeWeese always argued from this strange perspective that made it impossible to answer him. This was
the perspective that gave Phaedrus the idea DeWeese
was concealing something from him. It wasn't until the
very end of his stay in Bozeman that he thought he
saw, in his own analytic and methodical way, what that
perspective was.

At the park entrance we stop and pay a man in a
Smokey Bear hat. He hands us a one-day pass in return. Ahead I see an elderly tourist take a movie of us,
then smile. From under his shorts protrude white legs
into street stockings and shoes. His wife, who watches

approvingly, has identical legs. I wave to them as we go by and they wave back. It's a moment that will be preserved on film for years.

Phaedrus despised this park without knowing exactly why—because he hadn't discovered it himself, perhaps, but probably not. Something else. The guided-tour attitude of the rangers angered him. The Bronx Zoo attitudes of the tourists disgusted him even more. Such a difference from the high country all around. It seemed an enormous museum with exhibits carefully manicured to give the illusion of reality, but nicely chained off so that children would not injure them. People entered the park and became polite and cozy and fakey to each other because the atmosphere of the park made them that way. In the entire time he had lived within a hundred miles of it he had visited it only once or twice.

But this is getting out of sequence. There's a span of about ten years missing. He didn't jump from Immanuel Kant to Bozeman, Montana. During this span of ten years he lived in India for a long time studying Oriental philosophy at Benares Hindu University.

As far as I know he didn't learn any occult secrets there. Nothing much happened at all except exposures. He listened to philosophers, visited religious persons, absorbed and thought and then absorbed and thought some more, and that was about all. All his letters show is an enormous confusion of contradictions and incongruities and divergences and exceptions to any rule he formulated about the things he observed. He'd entered India an empirical scientist, and he left India an empirical scientist, not much wiser than he had been when he'd come. However, he'd been exposed to a lot and had acquired a kind of latent image that appeared in conjunction with many other latent images later on.

Some of these latencies should be summarized because they become important later on. He became aware that the doctrinal differences among Hinduism and Buddhism and Taoism are not anywhere near as important as doctrinal differences among Christianity and Islam and Judaism. Holy wars are not fought over them because verbalized statements about reality are never presumed to be reality itself.

In all of the Oriental religions great value is placed on the Sanskrit doctrine of *Tat tvam asi,* "Thou are that," which asserts that everything you think you are and everything you think you perceive are undivided. To realize fully this lack of division is to become enlightened.

Logic presumes a separation of subject from object; therefore logic is not final wisdom. The illusion of separation of subject from object is best removed by the elimination of physical activity, mental activity and emotional activity. There are many disciplines for this. One of the most important is the Sanskrit *dhyana,* mispronounced in Chinese as "Chan" and again mispronounced in Japanese as "Zen." Phaedrus never got involved in meditation because it made no sense to him. In his entire time in India "sense" was always logical consistency and he couldn't find any honest way to abandon this belief. That, I think, was creditable on his part.

But one day in the classroom the professor of philosophy was blithely expounding on the illusory nature of the world for what seemed the fiftieth time and Phaedrus raised his hand and asked coldly if it was believed that the atomic bombs that had dropped on Hiroshima and Nagasaki were illusory. The professor smiled and said yes. That was the end of the exchange.

Within the traditions of Indian philosophy that answer may have been correct, but for Phaedrus and for anyone else who reads newspapers regularly and is concerned with such things as mass destruction of human beings that answer was hopelessly inadequate. He left the classroom, left India and gave up.

He returned to his Midwest, picked up a practical degree of journalism, married, lived in Nevada and Mexico, did odd jobs, worked as a journalist, a science writer and an industrial-advertising writer. He fathered two children, bought a farm and a riding horse and two cars and was starting to put on middle-aged weight. His pursuit of what has been called the ghost of reason had been given up. That's extremely important to understand. He had given up.

Because he'd given up, the surface of life was comfortable for him. He worked reasonably hard, was easy to get along with and, except for an occasional glimpse of inner emptiness shown in some short stories he wrote at the time, his days passed quite usually.

What started him up here into these mountains isn't certain. His wife seems not to know, but I'd guess it was perhaps some of those inner feelings of failure and the hope that somehow this might take him back on the track again. He had become much more mature, as if the abandonment of his inner goals had caused him somehow to age more quickly.

*   *   *

We exit from the park at Gardiner, where not much rain seems to fall, because the mountainsides show only grass and sage in the twilight. We decide to stay here for the night.

The town is on high banks on either side of a bridge over a river which rushes over smooth and clean boulders. Across the bridge they've already turned the lights on at the motel where we check in, but even in the artificial light coming from the windows I can see each cabin has been carefully surrounded by planted flowers, and so I step carefully to avoid them.

I notice things about the cabin too, which I point out to Chris. The windows are all double-hung and sash-weighted. The doors click shut without looseness. All the moldings are perfectly mitered. There's nothing arty about all this, it's just well done and, something tells me, is all done by one person.

When we return to the motel from the restaurant an elderly couple are sitting in a small garden outside the office enjoying the evening breeze. The man confirms that he's made all these cabins himself, and is so pleased it's been noticed that his wife, who sees this, invites us all to sit down.

We talk with no need to hurry. This is the oldest entrance to the park. It was used before there were any automobiles. They talk about changes that have taken place over the years, adding a dimension to what we see around us, and it builds to a kind of beautiful thing—this town, this couple and the years that have gone by here. Sylvia puts her hand on John's arm. I am conscious of the sounds of the river rushing past boulders below and a fragrance in the night wind. The woman, who knows all fragrances, says it is honeysuckle, and we are quiet for a while and I become pleasantly drowsy. Chris is almost asleep when we decide to turn in.

13

John and Sylvia eat their breakfast hot cakes and drink
their coffee, still caught up in the mood of last night,
but I'm finding it hard to get food down.

Today we should arrive at the school, the place
where an enormous coalescence of things occurred,
and I'm already feeling tense.

I remember reading once about an archeological ex-
cavation in the Near East, learning about the archeologist's feelings when he opened the forgotten tombs for
the first time in thousands of years. Now I feel like
some archeologist myself.

The sagebrush down the canyon now toward Liv-
ingston is like sagebrush you see all the way from here
into Mexico.

This morning sunlight is the same as yesterday's except warmer and softer now that we're at a lower altitude again.

There is nothing unusual.

It's just this archeological feeling that the calmness of the surroundings conceals things. A haunted place.

I really don't want to go there. I'd just as soon turn around and go back.

Just tension, I guess.

It fits one of the fragments of this memory, in which many mornings the tension was so intense he would throw up everything before he got to his first classroom. He loathed appearing before classrooms of students and talking. It was a complete violation of his whole lone, isolated way of life, and what he experienced was intense stage fright, except that it never showed on him as stage fright, but rather as a terrific *intensity* about everything he did. Students had told his wife it was just like electricity in the air. The moment he entered the classroom all eyes turned on him and followed him as he walked to the front of the room. All conversation died to a hush and remained at a hush even though it was several minutes, often, before the class started. Throughout the hour the eyes never strayed from him.

He became much talked about, a controversial figure. The majority of students avoided his sections like the Black Death. They had heard too many stories.

The school was what could euphemistically be called a "teaching college." At a teaching college you teach and you teach and you teach with no time for research, no time for contemplation, no time for participation in outside affairs. Just teach and teach and teach until your mind grows dull and your creativity vanishes and you become an automaton saying the same dull things over and over to endless waves of innocent students who cannot understand why you are so dull, lose respect and fan this disrespect out into the community. The reason you teach and you teach and you teach is that this is a very clever way of running a college on the cheap while giving a false appearance of genuine education.

Yet despite this he called the school by a name that didn't make much sense, in fact sounded a little ludicrous in view of its actual nature. But the name had great meaning to him, and he stuck to it and he felt, before he left, that he had rammed it into a few minds sufficiently hard to make it stick. He called it a "Church of Reason," and much of the puzzlement people had about him could have ended if they'd understood what he meant by this.

The state of Montana at this time was undergoing an outbreak of ultra-right-wing politics like that which occurred in Dallas, Texas, just prior to President Kennedy's assassination. A nationally known professor from the University of Montana at Missoula was prohibited from speaking on campus on the grounds that it would "stir up trouble." Professors were told that all public statements must be cleared through the college public-relations office before they could be made.

Academic standards were demolished. The legislature had previously prohibited the school from refusing entry to any student over twenty-one whether he had a high-school diploma or not. Now the legislature had passed a law fining the college eight thousand dollars for every student who failed, virtually an order to pass every student.

The newly elected governor was trying to fire the college president for both personal and political reasons. The college president was not only a personal enemy, he was a Democrat, and the governor was no ordinary Republican. His campaign manager doubled as state coordinator for the John Birch Society. This was the same governor who supplied the list of fifty subversives we heard about a few days ago.

Now, as part of this vendetta, funds to the college were being cut. The college president had passed on an unusually large part of the cut to the English department, of which Phaedrus was a member, and whose members had been quite vocal on issues of academic freedom.

Phaedrus had given up, was exchanging letters with the Northwest Regional Accrediting Association to see if they could help prevent these violations of accreditation requirements. In addition to this private correspondence he had publicly called for an investigation of the entire school situation.

At this point some students in one of his classes had asked Phaedrus, bitterly, if his efforts to stop accreditation meant he was trying to prevent them from getting an education.

Phaedrus said no.

Then one student, apparently a partisan of the governor, said angrily that the legislature would prevent the school from losing its accreditation.

Phaedrus asked how.

The student said they would post police to prevent it.

Phaedrus pondered this for a while, then realized the enormity of the student's misconception of what accreditation was all about.

That night, for the next day's lecture, he wrote out his defense of what he was doing. This was the Church of Reason lecture, which, in contrast to his usual sketchy lecture notes, was very long and very carefully elaborated.

It began with reference to a newspaper article about a country church building with an electric beer sign hanging right over the front entrance. The building had been sold and was being used as a bar. One can guess that some classroom laughter started at this point. The college was well-known for drunken partying and the image vaguely fit. The article said a number of people had complained to the church officials about it. It had been a Catholic church, and the priest who had been delegated to respond to the criticism had sounded quite irritated about the whole thing. To him it had revealed an incredible ignorance of what a church really was. Did they think that bricks and boards and glass constituted a church? Or the shape of the roof? Here, posing as piety was an example of the very materialism the church opposed. The building in question was not holy ground. It had been desanctified. That was the end of it. The beer sign resided over a bar, not a church, and those who couldn't tell the difference were simply revealing something about themselves.

Phaedrus said the same confusion existed about the University and that was why loss of accreditation was hard to understand. The real University is not a material object. It is not a group of buildings that can be defended by police. He explained that when a college lost its accreditation, nobody came and shut down the school. There were no legal penalties, no fines, no jail sentences. Classes did not stop. Everything went on just as before. Students got the same education they would if the school didn't lose its accreditation. All that would happen, Phaedrus said, would simply be an official recognition of a condition that already existed. It would be similar to excommunication. What would happen is that the *real* University, which no legislature can dictate to and which can never be identified by any location of bricks or boards or glass, would simply declare that this place was no longer "holy ground." The real University would vanish from it, and all that would be left was the bricks and the books and the material manifestation.

It must have been a strange concept to all of the students, and I can imagine him waiting for a long time for it to sink in, and perhaps then waiting for the question, What do you think the real University is?

His notes, in response to this question, state the following:

The real University, he said, has no specific location. It owns no property, pays no salaries and receives no material dues. The real University is a state of mind. It is that great heritage of rational thought that has been brought down to us through the centuries and which does not exist at any specific location. It's a state of mind which is regenerated throughout the centuries by a body of people who traditionally carry the title of professor, but even that title is not part of the real University. The real University is nothing less than the continuing body of reason itself.

In addition to this state of mind, "reason," there's a legal entity which is unfortunately called by the same name but which is quite another thing. This is a nonprofit corporation, a branch of the state with a specific address. It owns property, is capable of paying salaries, of receiving money and of responding to legislative pressures in the process.

But this second university, the legal corporation, cannot teach, does not generate new knowledge or evaluate ideas. It is not the real University at all. It is just a church building, the setting, the location at which conditions have been made favorable for the real church to exist.

Confusion continually occurs in people who fail to see this difference, he said, and think that control of the church buildings implies control of the church. They see professors as employees of the second university who should abandon reason when told to and take orders with no backtalk, the same way employees do in other corporations.

They see the second university, but fail to see the first.

I remember reading this for the first time and remarking about the analytic craftsmanship displayed. He avoided splitting the University into fields or departments and dealing with the results of that analysis. He also avoided the traditional split into students, faculty and administration. When you split it either of those ways you get a lot of dull stuff that doesn't really tell you much you can't get out of the official school bulletin. But Phaedrus split it between "the church" and "the location," and once this cleavage is made the same rather dull and imponderable institution seen in the bulletin suddenly is seen with a degree of clarity that wasn't previously available. On the basis of this cleavage he provided explanations for a number of puzzling but normal aspects of University life.

After these explanations he returned to the analogy of the religious church. The citizens who build such a church and pay for it probably have in mind that they're doing this for the community. A good sermon can put the parishioners in a right frame of mind for the coming week. Sunday school will help the children grow up right. The minister who delivers the sermon and directs the Sunday school understands these goals and normally goes along with them, but he also knows that *his* primary goals are not to serve the community. His primary goal is always to serve God. Normally there's no conflict but occasionally one creeps in when trustees oppose the minister's sermons and threaten reduction of funds. That happens.

A true minister, in such situations, must act as though he'd never heard the threats. His primary goal isn't to serve the members of the community, but always God.

The primary goal of the Church of Reason, Phaedrus said, is always Socrates' old goal of truth, in its ever-changing forms, as it's revealed by the process of rationality. Everything else is subordinate to that. Normally this goal is in no conflict with the location goal of improving the citizenry, but on occasion some conflict arises, as in the case of Socrates himself. It arises when trustees and legislators who've contributed large amounts of time and money to the location take points of view in opposition to the professors' lectures or public statements. They can then lean on the administration by threatening to cut off funds if the professors don't say what they want to hear. That happens too.

True churchmen in such situations must act as though they had never heard these threats. Their primary goal never is to serve the community ahead of everything else. Their primary goal is to serve, through reason, the goal of truth.

That was what he meant by the Church of Reason. There was no question but that it was a concept that was deeply felt by him. He was regarded as something of a troublemaker but was never censured for it in any proportion to the amount of trouble he made. What saved him from the wrath of everyone around him was partly an unwillingness to give any support to the enemies of the college, but also partly a begrudging understanding that all of his troublemaking was ultimately motivated by a mandate they were never free from themselves: the mandate to speak the rational truth.

The lecture notes explain almost all of why he acted the way he did, but leave one thing unexplained—his fanatic intensity. One can believe in the truth and in the process of reason to discover it and in resistance to state legislatures, but why burn one's self out, day after day, over it?

The psychological explanations that have been made to me seem inadequate. Stage fright can't sustain that kind of effort month after month. Neither does another explanation sound right, that he was trying to redeem himself for his earlier failure. There is no evidence anywhere that he ever thought of his expulsion from the university as a failure, just an enigma. The explanation I've come to arises from the discrepancy between his lack of faith in scientific reason in the laboratory and his fanatic faith expressed in the Church of Reason lecture. I was thinking about the discrepancy one day and it suddenly came to me that it wasn't a discrepancy at all. His lack of faith in reason was *why* he was so fanatically dedicated to it.

You are never dedicated to something you have

complete confidence in. No one is fanatically shouting that the sun is going to rise tomorrow. They *know* it's going to rise tomorrow. When people are fanatically dedicated to political or religious faiths or any other kinds of dogmas or goals, it's always because these dogmas or goals are in doubt.

The militancy of the Jesuits he somewhat resembled is a case in point. Historically their zeal stems not from the strength of the Catholic Church but from its weakness in the face of the Reformation. It was Phaedrus' *lack* of faith in reason that made him such a fanatic teacher. That makes more sense. And it makes a lot of sense out of the things that followed.

That's probably why he felt such a deep kinship with so many failing students in the back rows of his classrooms. The contemptuous looks on their faces reflected the same feelings he had toward the whole rational, intellectual process. The only difference was that they were contemptuous because they didn't understand it. He was contemptuous because he did. Because they didn't understand it they had no solution but to fail and for the rest of their lives remember the experience with bitterness. He on the other hand felt fanatically obliged to do something about it. That was why his Church of Reason lecture was so carefully prepared. He was telling them you have to have faith in reason because there isn't anything else. But it was a faith he didn't have himself.

It must always be remembered that this was the nineteen-fifties, not the nineteen-seventies. There were rumblings from the beatniks and early hippies at this time about "the system" and the square intellectualism that supported it, but hardly anyone guessed how deeply the whole edifice would be brought into doubt. So here was Phaedrus, fanatically defending an institution, the Church of Reason, that no one, no one certainly in Bozeman, Montana, had any cause to doubt. A pre-Reformation Loyola. A militant reassuring everyone the sun would rise tomorrow, when no one was worried. They just wondered about *him.*

But now, with the most tumultuous decade of the century between him and ourselves, a decade in which reason has been assailed and assaulted beyond the wildest beliefs of the fifties, I think that in this Chautauqua based on his discoveries we can understand a little better what he was talking about . . . a solution for it all . . . if only that were true . . . so much of it's lost there's no way of knowing.

Maybe that's why I feel like an archeologist. And have such a tension about it. I have only these fragments of memory, and pieces of things people tell me, and I keep wondering as we get closer if some tombs are better left shut.

Chris, sitting behind me, suddenly comes to mind, and I wonder how much he knows, how much he remembers.

We reach an intersection where the road from the park joins the main east-west highway, stop and turn on to it. From here we go over a low pass and into Bozeman itself. The road goes up now, heading west, and suddenly I'm looking forward to what's ahead.

14

We ride down out of the pass onto a small green plain.
To the immediate south we can see pine-forested
mountains that still have last winter's snow on the
peaks. In all other directions appear lower mountains,
more in the distance, but just as clear and sharp. This
picture-postcard scenery vaguely fits memory but not
definitely. This interstate freeway we are on must not
have existed then.

The statement "To travel is better than to arrive"
comes back to mind again and stays. We have been
traveling and now we will arrive. For me a period of
depression comes on when I reach a temporary goal
like this and have to reorient myself toward another
one. In a day or two John and Sylvia must go back and
Chris and I must decide what we want to do next.
Everything has to be reorganized.

The main street of the town seems vaguely familiar
but there's a feeling of being a tourist now and I see the
shop signs are for me, the tourist, and not for people

who live here. This isn't really a small town. People are moving too fast and too independently of one another. It's one of these population-fifteen-to-thirty-thousand towns that isn't exactly a town, not exactly a city—not exactly anything really.

We eat lunch in a glass-and-chrome restaurant that brings no recall at all. It looks as though it's been built since he lived here and shows the same lack of self-identity seen on the main street.

I go to a phone book and look for Robert DeWeese's number but don't find it. I dial the operator but she's never heard of the party and can't tell me the number. I don't believe it! Were they just in his imagination? Her statement produces a panicky feeling that lasts for a moment, but then I remember their answer to my letter telling them we were coming and calm down. Imaginary people don't use the mails.

John suggests I try to call the art department or some friends. I smoke for a while and drink coffee, and when I'm relaxed again I do this and learn how to get there. It's not the technology that's scary. It's what it does to the relations between people, like callers and operators, that's scary.

From the town to the mountains across the valley floor must be less than ten miles, and we cross that distance now on dirt roads through rich green high alfalfa ready for cutting, so thick it looks difficult to walk through. The fields sweep outward and slightly upward to the base of the mountains where a much darker green of the pines rises suddenly up. That will be where the DeWeeses live. Where the light green and the dark green meet. The wind is full of the light-green new-mown-hay smells and livestock smells. At one point we pass through a cold bank of air where the smell changes to pine, but then are back in the warmth again. Sunlight and meadows and the close-looming mountain.

Just as we get to the pines, the gravel in the road becomes very deep. We slow down to first gear and ten miles an hour and I keep both feet off the pegs to kick the cycle upright again if it should mush into the gravel and start to go down. We round a corner and suddenly enter the pines and a very steep V canyon in the mountain, and there right beside the road is a large grey house with an enormous abstract iron sculpture attached to one side and beneath it sitting in a chair tipped back against the house surrounded by company is the living image of DeWeese himself with a can of beer in his hand, which waves to us. Right out of the old photographs.

I'm so busy keeping the machine up I can't take my hands off the grips and I wave a leg back instead. The living image of DeWeese himself grins as we pull up.

"You found it," he says. Relaxed smile. Happy eyes.

"It's been a long time," I say. I feel happy too, though strange at suddenly seeing the image move and talk.

We dismount and take off our riding gear and I see that the open porch deck he and his guests are on is unfinished and unweathered. DeWeese looks down from where it is only a few feet above the road on our side, but the V of the canyon slants so steeply that on the far side the ground descends fifteen feet below the deck. The stream itself appears another fifty feet down and away from the house, among trees and deep grass where a horse, partially hidden by the trees, grazes without looking up. Now we have to look high to see the sky. Surrounding us is the dark-green forest we watched as we approached.

"This is just *beautiful*!" Sylvia says.

The living image of DeWeese smiles down at her. "Thank you," he says, "I'm glad you like it." His tone is all here and now, completely relaxed. I realize that although this is the authentic image of DeWeese himself, it's also a brand-new person who's been renewing himself continually and I'm going to have to get to know him all over again.

We step up onto the deck. Between the floorboards it has spaces, like a grate. I can see the ground through them. With a "Well, I'm not quite sure how to do this" tone and smile, DeWeese makes introductions all around, but they're in one ear and out the other. I can never remember names. His guests are an art instructor from the school who has horn-rimmed glasses, and his wife, who smiles self-consciously. They must be new.

We talk for a while, DeWeese mainly explaining to them who I am, and then, from where the deck disappears around the corner of the house, suddenly comes Gennie DeWeese with a tray of beer cans. She is a painter too and, I'm suddenly aware, a quick comprehender and already there's a shared smile over the artistic economy of grabbing a can of beer instead of her hand, while she says, "Some neighbors just came over with a mess of trout for dinner. I'm so pleased." I try to think of something appropriate to say, but just nod.

We sit down, I in the sunlight, where it's difficult to distinguish details of the other side of the deck in the shade.

DeWeese looks at me, seems about to comment on my appearance, which is undoubtedly much different from what he remembers, but something deflects this and he turns to John instead and asks about the trip.

John explains that it's been just great, something he and Sylvia have needed for years.

Sylvia seconds this. "Just to be out in the open in all this space," she says.

"Lots of space in Montana," DeWeese says, a little wistfully. He and John and the art instructor become involved in get-acquainted talk about differences between Montana and Minnesota.

The horse grazes peacefully below us, and just beyond it the water sparkles in the creek. The talk has shifted to DeWeese's land here in the canyon, how long DeWeese has lived here and what art instruction at the college is like. John has a real gift for casual conversation like this that I've never had, so I just listen.

After a while the heat from the sun is so great I take off my sweater and open my shirt. Also to stop squinting I bring out some sunglasses and put them on. That's better, but it blanks out the shade so completely I can hardly see faces at all and leaves me feeling sort of visually detached from everything but the sun and the sunlit slopes of the canyon. I think to myself about unpacking but decide not to mention it. They know we're staying but just intuitively allow first things to happen first. First we relax, then we unpack. What's the hurry? The beer and sun begin to toast my head like a marshmallow. Very nice.

I don't know how much later I hear some comments about "the movie star here" come from John and I realize he is talking about me and my sunglasses. I look over the tops of them into the shade and make out that DeWeese and John and the art instructor are smiling at me. They must want me in the conversation, something about problems on the trip.

"They want to know what happens if something goes bad mechanically," John says.

I relate the whole story of the time Chris and I were in the rainstorm and the engine quit, which is a good story, but somewhat pointless, I realize as I'm telling it, as an answer to his question. The final line about being out of gas brings the expected groan.

"And I even told him to look," Chris says.

Both DeWeese and Gennie comment on Chris's size. He becomes self-conscious and glows a little. They ask about his mother and his brother and we both answer these questions as best we can.

The heat of the sun finally becomes too much for me and I shift my chair into the shade. The marshmallow feeling leaves in the sudden chill and after a few minutes I have to button up. Gennie notices and says, "As soon as the sun goes over the ridge up there it gets really cold."

The distance between the sun and the ridge is narrow. I'd judge that although it's only the middle of the afternoon, less than half an hour of direct sun remains. John asks about the mountains in the winter and he and DeWeese and the art instructor talk about this and about snowshoeing in the mountains. I could just sit here forever.

Sylvia and Gennie and the art instructor's wife talk about the house and soon Gennie invites them inside.

My thoughts drift to the statement about Chris

growing so fast and suddenly the feeling of the tomb comes on. I've heard only indirectly of the time Chris lived here, and yet to them it seems that he's hardly been gone. We live in entirely different time structures.

The conversation shifts onto what is current in art and music and theater and I'm surprised at how well John keeps up his end of the conversation. I'm not basically interested in what's new in these areas and he probably knows it and for that reason never talks about it to me. Just the reverse of the motorcycle maintenance situation. I wonder if I look as glassy-eyed now as he does when I talk about rods and pistons.

But what he and DeWeese really have in common is Chris and me, and a funny stickiness is developing here, ever since the movie-star comment. John's good-natured sarcasm toward his old drinking and cycling companion is chilling DeWeese slightly, causing reluctant respectful tones toward me from DeWeese. These seem to increase John's sarcasm in a self-stoking way and they both sense this and so they kind of veer away from me onto some subjects of agreement and then come back again but this stickiness develops and they veer away again onto another agreeable subject.

"Anyway," John says, "this character here told us we were in for a letdown when we came here, and we still haven't gotten over this 'letdown.' "

I laugh. I hadn't wanted to build him up to it. DeWeese smiles too. But then John turns to me and says, "Geez, you must have been *really* crazy, I mean really *nuts* to leave this place. I don't care what the college is like."

I see DeWeese look at him, shocked. Then angry. DeWeese looks at me and I wave it off. Some kind of impasse has developed but I don't know how to get around it. "It's a beautiful place," I say weakly.

DeWeese says defensively, "If you were here for a while you'd see another side to it." The instructor nods in agreement.

The impasse now produces its silence. It's an impossible one to reconcile. What John said wasn't unkind. He's kinder than anyone. What he knows and I know but DeWeese doesn't know is that the person they're both referring to isn't much these days. Just another middle-class, middle-aged person getting along. Worried mainly about Chris, but beyond that nothing special.

But what DeWeese and I know and the Sutherlands don't know is that there *was* someone, a person who lived here once, who was creatively on fire with a set of ideas no one had ever heard of before, but then something unexplained and wrong happened and DeWeese doesn't know how or why and neither do I. The reason for the impasse, the bad feeling, is that DeWeese thinks that person is here now. And there's no way I can tell him otherwise.

For a brief moment, way up at the top of the ridge, the sun diffuses through the trees and a halation of the light comes down to us. The halo expands, capturing everything in a sudden flash, and suddenly it catches me too.

"He saw too much," I say, still thinking about the impasse, but DeWeese looks puzzled and John doesn't register at all, and I realize the *non sequitur* too late. In the distance a single bird cries plaintively.

Now suddenly the sun is gone behind the mountain and the whole canyon is in dull shadow.

To myself I think how uncalled for that was. You don't make statements like that. You leave the hospital with the understanding that you don't.

Gennie appears with Sylvia and suggests we unpack. We agree and she shows us to our rooms. I see that my bed has a heavy quilt on it against the cold of the night. Beautiful room.

In three trips to the cycle and back I have everything transferred. Then I go to Chris's room to see what needs to be unpacked but he's cheerful and being grown-up and doesn't need help.

I look at him. "How do you like it here?"

He says, "Fine, but it isn't anything like the way you told about it last night."

"When?"

"Just before we went to sleep. In the cabin."

I don't know what he's referring to.

He adds, "You said it was lonely here."

"Why would I say *that*?"

"*I* don't know." My question frustrates him, so I leave it. He must have been dreaming.

When we come down to the living room I can smell the aroma from the frying trout in the kitchen. At one end of the room DeWeese is bent over the fireplace holding a match to some newspaper under the kindling. We watch him for a while.

"We use this fireplace all summer long," he says.

I reply, "I'm surprised it's this cold."

Chris says he's cold too. I send him back up for his sweater and mine as well.

"It's the evening wind," DeWeese says. "It sweeps down the canyon from up high where it's really cold."

The fire flares suddenly and then dies and then

flares again from an uneven draft. It must be windy, I think, and look through the huge windows that line one wall of the living room. Across the canyon in the dusk I see the sharp movement of the trees.

"But that's right," DeWeese says. "You know how cold it is up there. You used to spend all your time up there."

"It brings back memories," I say.

A single fragment comes to mind now of night winds all around a campfire, smaller than this one before us now, sheltered in the rock against the high wind because there are no trees. Next to the fire are cooking gear and backpacks to help give wind shelter, and a canteen filled with water gathered from the melting snow. The water had to be collected early because above the timberline the snow stops melting when the sun goes down.

DeWeese says, "You've changed a lot." He is looking at me searchingly. His expression seems to ask whether this is a forbidden topic or not, and he gathers from looking at me that it is. He adds, "I guess we all have."

I reply, "I'm not the same person at all," and this seems to put him a little more at ease. Were he aware of the literal truth of that, he'd be a lot *less* at ease. "A lot has happened," I say, "and some things have come up that have made it important to try to untangle them a little, in my own mind at least, and that's partly why I'm here."

He looks at me, expecting something more, but the art instructor and his wife appear by the fireside and we drop it.

"The wind sounds like there'll be a storm tonight," the instructor says.

"I don't think so," DeWeese says.

Chris returns with the sweaters and asks if there are any ghosts up in the canyon.

DeWeese looks at him with amusement. "No, but there are wolves," he says.

Chris thinks about this and asks, "What do *they* do?"

DeWeese says, "They make trouble for the ranchers." He frowns. "They kill the young calves and lambs."

"Do they chase people?"

"I've never heard of it," DeWeese says and then, seeing that this disappoints Chris, adds, "but they *could*."

At dinner the brook trout is accompanied by glasses of Bay county Chablis. We sit separately in chairs and sofas around the living room. One entire side of this room has the windows which would overlook the canyon, except that now it's dark outside and the glass reflects the light from the fireplace. The glow of the fire is matched by an inner glow from the wine and fish and we don't say much except murmurs of appreciation.

Sylvia murmurs to John to notice the large pots and vases around the room.

"I *was* noticing those," John says. "Fantastic."

"Those were made by Peter Voulkas," Sylvia says.

"Is *that* right?"

"He was a student of Mr. DeWeese."

"Oh, for Christ's sake! I almost kicked one of those over."

DeWeese laughs.

Later John mumbles something a few times, looks up and announces, "This does it . . . this just does the whole thing for us. . . . Now we can go back for another eight years on Twenty-six-forty-nine Colfax Avenue."

Sylvia says mournfully, "Let's not talk about that."

John looks at me for a moment. "I suppose anybody with friends who can provide an evening like this can't be *all* bad." He nods gravely. "I'm going to have to take back all those things I thought about you."

"*All* of them?" I ask.

"*Some,* anyway."

DeWeese and the instructor smile and some of the impasse goes away.

After dinner Jack and Wylla Barsness arrive. More living images. Jack is recorded in the tomb fragments as a good person who writes and teaches English at the college. Their arrival is followed by that of a sculptor from northern Montana who herds sheep for his income. I gather from the way DeWeese introduces him that I'm not supposed to have met him before.

DeWeese says he is trying to persuade the sculptor to join the faculty and I say, "I'll try to talk him out of it," and sit down next to him, but conversation is very sticky because the sculptor is extremely serious and suspicious, evidently because I'm not an artist. He acts like I'm a detective trying to get something on him, and it isn't until he discovers I do a lot of welding that I become okay. Motorcycle maintenance opens strange doors. He says he welds for some of the same reasons I do. After you pick up skill, welding gives a tremendous feeling of power and control over the metal. You can do anything. He brings out some photographs of things he has welded and these show beautiful birds and animals with flowing metal surface textures that are not like anything else.

Later I move over and talk with Jack and Wylla. Jack
is leaving to head an English department down in
Boise, Idaho. His attitudes toward the department here
seem guarded, but negative. They would be negative,
of course, or he wouldn't be leaving. I seem to remember now he was a fiction writer mainly, who
taught English, rather than a systematic scholar who
taught English. There was a continuing split in the department along these lines which in part gave rise to, or
at least accelerated the growth of, Phaedrus' wild set of
ideas which no one else had ever heard of, and Jack
was supportive of Phaedrus because, although he wasn't
sure he knew what Phaedrus was talking about, he saw
it was something a fiction writer could work with better than linguistic analysis. It's an old split. Like the
one between art and art history. One does it and the
other talks about how it's done and the talk about how
it's done never seems to match how one does it.

DeWeese brings over some instructions for assembly of an outdoor barbecue rotisserie which he wants
me to evaluate as a professional technical writer. He's
spent a whole afternoon trying to get the thing together
and he wants to see these instructions totally damned.

But as I read them they look like ordinary instructions to me and I'm at a loss to find anything wrong
with them. I don't want to say this, of course, so I hunt
hard for something to pick on. You can't really tell
whether a set of instructions are all right until you
check them against the device or procedure they describe, but I see a page separation that prevents reading
without flipping back and forth between the text and illustration—always a poor practice. I jump on this very

hard and DeWeese encourages every jump. Chris takes the instructions to see what I mean.

But while I'm jumping on this and describing some of the agonies of misinterpretation that bad cross-referencing can produce, I've a feeling that this isn't why DeWeese found them so hard to understand. It's just the lack of smoothness and continuity which threw him off. He's unable to comprehend things when they appear in the ugly, chopped-up, grotesque sentence style common to engineering and technical writing. Science works with chunks and bits and pieces of things with the continuity presumed, and DeWeese works only with the continuities of things with the chunks and bits and pieces presumed. What he really wants me to damn is the lack of artistic continuity, something an engineer couldn't care less about. It hangs up, really, on the classic-romantic split, like everything else about technology.

But Chris, meanwhile, takes the instructions and folds them around in a way I hadn't thought of so that the illustration sits there right next to the text. I double-take this, then triple-take it and feel like a movie-cartoon character who has just walked beyond the edge of a cliff but hasn't fallen yet because he hasn't realized his predicament. I nod, and there's silence, and then I realize my predicament, then a long laughter as I pound Chris on the top of the head all the way down to the bottom of the canyon. When the laughter subsides, I say, "Well, anyway . . ." but the laughter starts all over again.

"What I wanted to say," I finally get in, "is that I've a set of instructions at home which open up great realms for the improvement of technical writing. They begin, 'Assembly of Japanese bicycle require great peace of mind.' "

This produces more laughter, but Sylvia and Gennie and the sculptor give sharp looks of recognition.

"That's a *good* instruction," the sculptor says. Gennie nods too.

"That's kind of why I saved it," I say. "At first I laughed because of memories of bicycles I'd put together and, of course, the unintended slur on Japanese manufacture. But there's a lot of wisdom in that statement."

John looks at me apprehensively. I look at him with equal apprehension. We both laugh. He says, "The professor will now expound."

"Peace of mind isn't at all superficial, really," I expound. "It's the whole thing. That which produces it is good maintenance; that which disturbs it is poor maintenance. What we call workability of the machine is just an objectification of this peace of mind. The ultimate test's always your own serenity. If you don't have this when you start and maintain it while you're working you're likely to build your personal problems right into the machine itself."

They just look at me, thinking about this.

"It's an unconventional concept," I say, "but conventional reason bears it out. The material object of observation, the bicycle or rotisserie, can't be right or wrong. Molecules are molecules. They don't have any ethical codes to follow except those people give them. The test of the machine is the satisfaction it gives you. There isn't any other test. If the machine produces tranquillity it's right. If it disturbs you it's wrong until either the machine or your mind is changed. The test of the machine's always your own mind. There isn't any other test."

DeWeese asks, "What if the machine is wrong and I feel peaceful about it?"

Laughter.

I reply, "That's self-contradictory. If you *really* don't care you aren't going to *know* it's wrong. The thought'll never occur to you. The act of pronouncing it wrong's a form of caring."

I add, "What's more common is that you feel un-peaceful even if it's right, and I think that's the actual case here. In this case, if you're worried, it isn't right. That means it isn't checked out thoroughly enough. In any industrial situation a machine that isn't checked out is a 'down' machine and can't be used even though it may work perfectly. Your worry about the rotisserie is the same thing. You haven't completed the ultimate requirement of achieving peace of mind, because you feel these instructions were too complicated and you may not have understood them correctly."

DeWeese asks, "Well, how would you change them so I would get this peace of mind?"

"That would require a lot more study than I've just given them now. The whole thing goes very deep. These rotisserie instructions begin and end exclusively with the machine. But the kind of approach I'm think-ing about doesn't cut it off so narrowly. What's really angering about instructions of this sort is that they imply there's only one way to put this rotisserie together—*their* way. And that presumption wipes out all the creativity. Actually there are hundreds of ways to put the rotisserie together and when they make you follow just one way without showing you the overall problem the instructions become hard to follow in such a way as not to make mistakes. You lose feeling for the work. And not only that, it's very unlikely that they've told you the best way."

"But they're from the *factory*," John says.

"*I'm* from the factory, too," I say, "and I *know* how instructions like this are put together. You go out on the assembly line with a tape recorder and the foreman sends you to talk to the guy he needs least, the biggest goof-off he's got, and whatever he tells you—that's the instructions. The next guy might have told you something completely different and probably better, but he's too busy."

They all look surprised.

"I might have known," DeWeese says.

"It's the format," I say. "No writer can buck it. Technology presumes there's just one right way to do things and there never is. And when you presume there's just one right way to do things, of *course* the instructions begin and end exclusively with the rotisserie. But if you have to choose among an infinite number of ways to put it together then the relation of the machine to you, and the relation of the machine and you to the rest of the world, has to be considered, because the selection from among many choices, the *art* of the work is just as dependent upon your own mind and spirit as it is upon the material of the machine. That's why you need the peace of mind.

"Actually this idea isn't so strange," I continue. "Sometime look at a novice workman or a bad workman and compare his expression with that of a craftsman whose work you know is excellent and you'll see the difference. The craftsman isn't ever following a single line of instruction. He's making decisions as he goes along. For that reason he'll be absorbed and attentive to what he's doing even though he doesn't deliberately contrive this. His motions and the machine are in a kind of harmony. He isn't following any set of written instructions because the nature of the material at hand determines his thoughts and motions, which simultaneously change the nature of the material at hand. The material and his thoughts are changing together in a progression of changes until his mind's at rest at the same time the material's right."

"Sounds like art," the instructor says.

"Well, it *is* art," I say. "This divorce of art from technology is completely unnatural. It's just that it's gone on so long you have to be an archeologist to find out where the two separated. Rotisserie assembly is actually a long-lost branch of sculpture, so divorced from its roots by centuries of intellectual wrong turns that just to associate the two sounds ludicrous."

They're not sure whether I'm kidding or not.

"You mean," DeWeese asks, "that when I was putting this rotisserie together I was actually sculpting it?"

"Sure."

He goes over this in his mind, smiling more and more. "I wish I'd known that," he says. Laughter follows.

Chris says he doesn't understand what I'm saying.

"That's all right, Chris," Jack Barsness says. "We don't either." More laughter.

"I think I'll just stay with ordinary sculpture," the sculptor says.

"I think I'll just stick to painting," DeWeese says.

"I think I'll just stick to drumming," John says.

Chris asks, "What are you going to stick to?"

"Mah guns, boy, mah guns," I tell him. "That's the Code of the West."

They all laugh hard at this, and my speechifying seems forgiven. When you've got a Chautauqua in your head, it's extremely hard not to inflict it on innocent people.

The conversation breaks up into groups and I spend the rest of the party talking to Jack and Wylla about developments in the English department.

After the party is over and the Sutherlands and Chris have gone to bed, DeWeese recalls my lecture, however. He says seriously, "What you said about the rotisserie instructions was interesting."

Gennie adds, also seriously, "It sounded like you had been thinking about it for a long time."

"I've been thinking about concepts that underlie it for twenty years," I say.

Beyond the chair in front of me, sparks fly up the chimney, drawn by the wind outside, now stronger than before.

I add, almost to myself, "You look at where you're going and where you are and it never makes sense, but then you look back at where you've been and a pattern seems to emerge. And if you project forward from that pattern, then sometimes you can come up with something.

"All that talk about technology and art is part of a pattern that seems to have emerged from my own life. It represents a transcendence from something I think a lot of others may be trying to transcend."

"What's that?"

"Well, it isn't just art and technology. It's a kind of a noncoalescence between reason and feeling. What's wrong with technology is that it's not connected in any real way with matters of the spirit and of the heart. And so it does blind, ugly things quite by accident and gets hated for that. People haven't paid much attention to this before because the big concern has been with food, clothing and shelter for everyone and technology has provided these.

"But now where these are assured, the ugliness is being noticed more and more and people are asking if we must always suffer spiritually and esthetically in order to satisfy material needs. Lately it's become almost a national crisis—antipollution drives, antitechnological communes and styles of life, and all that."

Both DeWeese and Gennie have understood all this for so long there's no need for comment, so I add, "What's emerging from the pattern of my own life is the belief that the crisis is being caused by the inadequacy of existing forms of thought to cope with the situation. It can't be solved by rational means because the rationality itself is the source of the problem. The only ones who're solving it are solving it at a personal level by abandoning 'square' rationality altogether and going by feelings alone. Like John and Sylvia here. And millions of others like them. And that seems like a wrong direction too. So I guess what I'm trying to say is that the solution to the problem isn't that you abandon rationality but that you expand the nature of rationality so that it's capable of coming up with a solution."

"I guess I don't know what you mean," Gennie says.

"Well, it's quite a bootstrap operation. It's analogous

to the kind of hang-up Sir Isaac Newton had when he wanted to solve problems of instantaneous rates of change. It was unreasonable in his time to think of anything changing within a zero amount of time. Yet it's almost necessary mathematically to work with other zero quantities, such as points in space and time that no one thought were unreasonable at all, although there was no real difference. So what Newton did was say, in effect, 'We're going to *presume* there's such a thing as instantaneous change, and see if we can find ways of determining what it is in various applications.' The result of this presumption is the branch of mathematics known as the calculus, which every engineer uses today. Newton *invented* a new form of reason. He expanded reason to handle infinitesimal changes and I think what is needed now is a similar expansion of reason to handle technological ugliness. The trouble is that the expansion has to be made at the roots, not at the branches, and that's what makes it hard to see.

"We're living in topsy-turvy times, and I think that what causes the topsy-turvy feeling is inadequacy of old forms of thought to deal with new experiences. I've heard it said that the only real learning results from hang-ups, where instead of expanding the branches of what you already know, you have to stop and drift laterally for a while until you come across something that allows you to expand the roots of what you already know. Everyone's familiar with that. I think the same thing occurs with whole civilizations when expansion's needed at the roots.

"You look back at the last three thousand years and with hindsight you think you see neat patterns and chains of cause and effect that have made things the way they are. But if you go back to original sources, the literature of any particular era, you find that these causes were never apparent at the time they were supposed to be operating. During periods of root expansion things have always looked as confused and topsy-turvy and purposeless as they do now. The whole Renaissance is supposed to have resulted from the topsy-turvy feeling caused by Columbus' discovery of a new world. It just shook people up. The topsy-turviness of that time is recorded everywhere. There was nothing in the flat-earth views of the Old and New Testaments that predicted it. Yet people couldn't deny it. The only way they could assimilate it was to abandon the entire medieval outlook and enter into a new expansion of reason.

"Columbus has become such a schoolbook stereotype it's almost impossible to imagine him as a living human being anymore. But if you really try to hold back your present knowledge about the consequences of his trip and project yourself into his situation, then sometimes you can begin to see that our present moon exploration must be like a tea party compared to what he went through. Moon exploration doesn't involve real root expansions of thought. We've no reason to doubt that existing forms of thought are adequate to handle it. It's really just a branch extension of what Columbus did. A really new exploration, one that would look to us today the way the world looked to Columbus, would have to be in an entirely new direction."

"Like what?"

"Like into realms beyond reason. I think present-day reason is an analogue of the flat earth of the medieval period. If you go too far beyond it you're presumed to fall off, into insanity. And people are very much afraid of that. I think this fear of insanity is comparable to the fear people once had of falling off the edge of the world. Or the fear of heretics. There's a very close analogue there.

"But what's happening is that each year our old flat earth of conventional reason becomes less and less adequate to handle the experiences we have and this is creating widespread feelings of topsy-turviness. As a result we're getting more and more people in irrational areas of thought—occultism, mysticism, drug changes and the like—because they feel the inadequacy of classical reason to handle what they know are real experiences."

"I'm not sure what you mean by *classical* reason."

"Analytic reason, dialectic reason. Reason which at the University is sometimes considered to be the whole of understanding. You've never *had* to understand it really. It's always been completely bankrupt with regard to abstract art. Nonrepresentative art is one of the root experiences I'm talking about. Some people still condemn it because it doesn't make 'sense.' But what's really wrong is not the art but the 'sense,' the classical reason, which can't grasp it. People keep looking for branch extensions of reason that will cover art's more recent occurrences, but the answers aren't *in* the branches, they're at the roots."

A rush of wind comes furiously now, down from the mountaintop. "The ancient Greeks," I say, "who were the inventors of classical reason, knew better than to use it exclusively to foretell the future. They listened to the wind and predicted the future from that. That sounds insane now. But why should the inventors of reason sound insane?"

DeWeese squints. "How could they tell the future from the wind?"

"I don't know, maybe the same way a painter can tell the future of his painting by staring at the canvas. Our whole system of knowledge stems from their results. We've yet to understand the methods that produced these results."

I think for a while, then say, "When I was last here, did I talk much about the Church of Reason?"

"Yes, you talked a lot about that."

"Did I ever talk about an individual named Phaedrus?"

"No."

"Who was he?" Gennie asks.

"He was an ancient Greek . . . a rhetorician . . . a 'composition major' of his time. He was one of those present when reason was being invented."

"You never talked about that, I don't think."

"That must have come later. The rhetoricians of ancient Greece were the first teachers in the history of the Western world. Plato vilified them in all his works to grind an axe of his own and since what we know about them is almost entirely from Plato they're unique in that they've stood condemned throughout history without ever having their side of the story told. The Church of Reason that I talked about was founded on their graves. It's supported today by their graves. And when you dig deep into its foundations you come across ghosts."

I look at my watch. It's after two. "It's a long story," I say.

"You should write all this down," Gennie says.

I nod in agreement. "I'm thinking about a series of lecture-essays—a sort of Chautauqua. I've been trying to work them out in my mind as we rode out here . . . which is probably why I sound so primed on all this stuff. It's all so huge and difficult. Like trying to travel through these mountains on foot.

"The trouble is that essays always have to sound like God talking for eternity, and that isn't the way it ever is. People should see that it's never anything other than just one person talking from one place in time and space and circumstance. It's never been anything else, ever, but you can't get that across in an essay."

"You should do it anyway," Gennie says. "Without trying to get it perfect."

"I suppose," I say.

DeWeese asks, "Does this tie in with what you were doing on 'Quality'?"

"It's the direct result of it," I say.

I remember something and look at DeWeese. "Didn't you advise me to drop it?"

"I said no one had ever succeeded in doing what you were trying to do."

"Do you think it's possible?"

"I don't know. Who knows?" His expression is really concerned. "A lot of people are listening better these days. Particularly the kids. They're really listening . . . and not just *at* you—*to* you . . . to *you*. It makes all the difference."

The wind coming down from the snowfields up above sounds for a long time throughout the house. It grows loud and high as if in hope of sweeping the whole house, all of us, away into nothing, leaving the canyon as it once was, but the house stands and the wind dies away again, defeated. Then it comes back, feinting a light blow from the far side, then suddenly a heavy gust from our side.

"I keep listening to the wind," I say.

I add, "I think when the Sutherlands have left, Chris and I should do some climbing up to where that wind starts. I think it's time he got a better look at that land."

"You can start from right here," DeWeese says, "and head back up the canyon. There's no road for seventy-five miles."

"Then this is where we'll start," I say.

Upstairs I'm glad to see the bed's heavy quilt again. It's become quite cold now and it'll be needed. I undress quickly and get way down deep under the quilt where it is warm, very warm, and think for a long time about snowfields and winds and Christopher Columbus.

15

For two days John and Sylvia and Chris and I loaf and talk and ride up to an old mining town and back, and then it's time for John and Sylvia to turn back home. We ride into Bozeman from the canyon now, together for the last time.

Up ahead Sylvia's turned around for the third time, evidently to see if we're all right. She's been very quiet the last two days. A glance from her yesterday seemed apprehensive, almost frightened. She worries too much about Chris and me.

At a bar in Bozeman we have one last round of beer, and I discuss routes back with John. Then we say perfunctory things about how good it's all been and how we'll see each other soon, and this is suddenly very sad to have to talk like this—like casual acquaintances.

Out in the street again Sylvia turns to me and Chris, pauses, and then says, "It'll be all right with you. There's nothing to worry about."

"Of course," I say.

Again that same frightened glance.

John has the motorcycle started and waits for her. "I believe you," I say.

She turns, gets on and with John watches oncoming traffic for an opportunity to pull out. "I'll see you," I say.

She looks at us again, expressionless this time. John finds his opportunity and enters into the traffic lane. Then Sylvia waves, as if in a movie. Chris and I wave back. Their motorcycle disappears in the heavy traffic of out-of-state cars, which I watch for a long time.

I look at Chris and he looks at me. He says nothing.

We spend the morning sitting at first on a park bench marked SENIOR CITIZENS ONLY, then get food and at a filling station change the tire and replace the chain adjuster link. The link has to be remachined to fit and so we wait and walk for a while, back away from the main street. We come to a church and sit down on the lawn in front of it. Chris lies back on the grass and covers his eyes with his jacket.

"You tired?" I ask him.

"No."

Between here and the edge of the mountains to the north, heat waves shimmer the air. A transparent-winged bug sets down from the heat on a stalk of grass by Chris's foot. I watch it flex its wings, feeling lazier every minute. I lie back to go to sleep, but don't. Instead a restless feeling hits. I get up.

"Let's walk for a while," I say.

"Where?"

"Toward the school."

"All right."

We walk under shady trees on very neat sidewalks

past neat houses. The avenues provide many small surprises of recognition. Heavy recall. He's walked through these streets many times. Lectures. He prepared his lectures in the peripatetic manner, using these streets as his academy.

The subject he'd been brought here to teach was rhetoric, writing, the second of the three R's. He was to teach some advanced courses in technical writing and some sections of freshman English.

"Do you remember this street?" I ask Chris.

He looks around and says, "We used to ride in the car to look for you." He points across the street. "I remember that house with the funny roof. . . . Whoever saw you first would get a nickel. And then we'd stop and let you in the back of the car and you wouldn't even talk to us."

"I was thinking hard then."

"That's what Mom said."

He *was* thinking hard. The crushing teaching load was bad enough, but what for him was far worse was that he understood in his precise analytic way that the subject he was teaching was undoubtedly the most unprecise, unanalytic, amorphous area in the entire Church of Reason. That's why he was thinking so hard. To a methodical, laboratory-trained mind, rhetoric is just completely hopeless. It's like a huge Sargasso Sea of stagnated logic.

What you're supposed to do in most freshman-rhetoric courses is to read a little essay or short story, discuss how the writer has done certain little things to achieve certain little effects, and then have the students write an imitative little essay or short story to see if they can do the same little things. He tried this over and over again but it never jelled. The students seldom achieved anything, as a result of this calculated mimicry, that was remotely close to the models he'd given them. More often their writing got worse. It seemed as though every rule he honestly tried to discover with them and learn with them was so full of exceptions and contradictions and qualifications and confusions that he wished he'd never come across the rule in the first place.

A student would always ask how the rule would apply in a certain special circumstance. Phaedrus would then have the choice of trying to fake through a made-up explanation of how it worked, or follow the selfless route and say what he really thought. And what he really thought was that the rule was *pasted on* to the writing after the writing was all done. It was *post hoc,* after the fact, instead of prior to the fact. And he became convinced that all the writers the students were supposed to mimic wrote without rules, putting down whatever sounded right, then going back to see if it still sounded right and changing it if it didn't. There were some who apparently wrote with calculating premeditation because that's the way their product looked. But that seemed to him to be a very poor way to look. It had a certain syrup, as Gertrude Stein once said, but it didn't pour. But how're you to teach something that isn't premeditated? It was a seemingly impossible requirement. He just took the text and commented on it in an unpremeditated way and hoped the students would get something from that. It wasn't satisfactory.

There it is up ahead. Tension hits, the same stomach feeling, as we walk toward it.

"Do you remember that building?"

"That's where you used to teach . . . why are we going here?"

"I don't know. I just wanted to see it."

Not many people seem to be around. There wouldn't be, of course. Summer session is on now. Huge and strange gables over old dark-brown brick. A beautiful building, really. The only one that really seems to belong here. Old stone stairway up to the doors. Stairs cupped by wear from millions of footsteps.

"Why are we going inside?"

"Shh. Just don't say anything now."

I open the great heavy outside door and enter. Inside are more stairs, worn and wooden. They creak underfoot and smell of a hundred years of sweeping and waxing. Halfway up I stop and listen. There's no sound at all.

Chris whispers, "Why are we *here*?"

I just shake my head. I hear a car go by outside.

Chris whispers, "I don't like it here. It's *scary* in here."

"Go outside then," I say.

"You come too."

"Later."

"No, now." He looks at me and sees I'm staying. His look is so frightened I'm about to change my mind, but then suddenly his expression breaks and he turns and runs down the stairs and out the door before I can follow him.

The big heavy door closes down below, and I'm all alone here now. I listen for some sound. . . . Of whom? . . . Of *him*? . . . I listen for a long time. . . .

The floorboards have an eerie creak as I move down

the corridor and they are accompanied by an eerie thought that it *is* him. In this place he is the reality and I am the ghost. On one of the classroom doorknobs I see his hand rest for a moment, then slowly turn the knob, then push the door open.

The room inside is waiting, exactly as remembered, as if he were here now. He *is* here now. He's aware of everything I see. Everything jumps forth and vibrates with recall.

The long dark-green chalkboards on either side are flaked and in need of repair, just as they were. The chalk, never any chalk except little stubs in the trough, is still here. Beyond the blackboard are the windows and through them are the mountains he watched, meditatively, on days when the students were writing. He would sit by the radiator with a stub of chalk in one hand and stare out the window at the mountains, interrupted, occasionally, by a student who asked, "Do we have to do . . . ?" And he would turn and answer whatever thing it was and there was a oneness he had never known before. This was a place where he was *received*— as himself. Not as what he could be or should be but as himself. A place all receptive—listening. He gave everything to it. This wasn't one room, this was a thousand rooms, changing each day with the storms and snows and patterns of clouds on the mountains, with each class, and even with each student. No two hours were ever alike, and it was always a mystery to him what the next one would bring. . . .

My sense of time has been lost when I hear a creaking of steps in the hall. It becomes louder, then stops at the entrance to this classroom. The knob turns. The door opens. A woman looks in.

She has an aggressive face, as if she intended to catch someone here. She appears to be in her late twenties, is not very pretty. "I thought I saw someone," she says. "I thought . . ." She looks puzzled.

She comes inside the room and walks toward me. She looks at me more closely. Now the aggressive look vanishes, slowly changing to wonder. She looks astonished.

"Oh, my God," she says. "Is it *you*?"

I don't recognize her at all. Nothing.

She calls my name and I nod, Yes, it's me.

"You've come back."

I shake my head. "Just for these few minutes."

She continues to look until it becomes embarrassing. Now she becomes aware of this herself, and asks, "May I sit down for a moment?" The timid way she asks this indicates she may have been a student of his.

She sits down on one of the front-row chairs. Her hand, which bears no wedding ring, is trembling. I really *am* a ghost.

Now she becomes embarrassed herself. "How long are you staying? . . . No, I asked you that . . ."

I fill in, "I'm staying with Bob DeWeese for a few clays and then heading West. I had some time to spend in town and thought I'd see how the college looked."

"Oh," she says. "I'm glad you did. . . . It's changed . . . we've all changed . . . so much since you left. . . ."

There's another embarrassing pause.

"We heard you were in the hospital. . . ."

"Yes," I say.

There is more embarrassed silence. That she doesn't pursue it means she probably knows why. She hesitates some more, searches for something to say. This is getting hard to bear.

"Where are you teaching?" she finally asks.

"I'm not teaching anymore," I say. "I've stopped."

She looks incredulous. "You've *stopped*?" She frowns and looks at me again, as if to verify that she is really talking to the right person. "You can't do that."

"Yes, you can."

She shakes her head in disbelief. "Not *you*!"

"Yes."

"Why?"

"That's all over for me now. I'm doing other things."

I keep wondering who she is, and her expression looks equally baffled. "But that's just . . ." The sentence drops off. She tries again. "You're just being completely . . ." but this sentence fails too.

The next word is "crazy." But she has caught herself both times. She realizes something, bites her lip and looks mortified. I'd say something if I could, but there's no place to start.

I'm about to tell her I don't know her but she stands up and says, "I must go now." I think she sees I don't know her.

She goes to the door, says good-bye quickly and perfunctorily, and as it closes her footsteps go quickly, almost at a run, down the hall.

The outer door of the building closes and the classroom is as silent as before, except for a kind of psychic eddy current she has left behind. The room is completely modified by it. Now it contains only the backwash of her presence, and what it was I came here to see has vanished.

Good, I think, standing up again, I'm glad to have

visited this room but I don't think I'll ever want to see it again. I'd rather fix motorcycles, and one's waiting.

On the way out I open one more door, compulsively. There on the wall I see something which sends a spine-tingling feeling along my neck.

It's a painting. I've had no recollection of it but now I know he *bought* it and put it there. And suddenly I know it's not a painting, it's a print of a painting he ordered from New York and which DeWeese had frowned at because it was a print and prints are *of* art and not art themselves, a distinction he didn't recognize at the time. But the print, Feininger's "Church of the Minorites," had an appeal to him that was irrelevant to the art in that its subject, a kind of Gothic cathedral, created from semiabstract lines and planes and colors and shades, seemed to reflect his mind's vision of the Church of Reason and that was why he'd put it here. All this comes back now. This was his office. A *find*. *This* is the room I am looking for!

I step inside and an avalanche of memory, loosened by the jolt of the print, begins to come down. The light on the print comes from a miserable cramped window in the adjacent wall through which he looked out onto and across the valley onto the Madison Range and watched the storms come in and while watching this valley before me now through this window here, now . . . started the whole thing, the whole madness, right *here*! This is the exact spot!

And that door leads to Sarah's office. Sarah! Now it comes down! She came trotting by with her watering pot between those two doors, going from the corridor to her office, and she said, "I hope you are teaching Quality to your students." This in a la-de-da, singsong voice of a lady in her final year before retirement about to water her plants. That was the moment it all started. That was the seed crystal.

Seed crystal. A powerful fragment of memory comes back now. The laboratory. Organic chemistry. He was working with an extremely supersaturated solution when something similar had happened.

A supersaturated solution is one in which the saturation point, at which no more material will dissolve, has been exceeded. This can occur because the saturation point becomes higher as the temperature of the solution is increased. When you dissolve the material at a high temperature and then cool the solution, the material sometimes doesn't crystallize out because the molecules don't know how. They require something to get them started, a seed crystal, or a grain of dust or even a sudden scratch or tap on the surrounding glass.

He walked to the water tap to cool the solution but never got there. Before his eyes, as he walked, he saw a star of crystalline material in the solution appear and then grow suddenly and radiantly until it filled the entire vessel. He *saw* it grow. Where before was only clear liquid there was now a mass so solid he could turn the vessel upside down and nothing would come out.

The one sentence "I hope you are teaching Quality to your students" was said to him, and within a matter of a few months, growing so fast you could almost *see* it grow, came an enormous, intricate, highly structured mass of thought, formed as if by magic.

I don't know what he replied to her when she said this. Probably nothing. She would be back and forth behind his chair many times each day to get to and from her office. Sometimes she stopped with a word or two of apology about the interruption, sometimes with a fragment of news, and he was accustomed to this as a part of office life. I know that she came by a second time and asked, "Are you *really* teaching Quality this quarter?" and he nodded and looked back from his chair for a second and said, "Definitely!" and she trotted on. He was working on lecture notes at the time and was in a state of complete depression about them.

What was depressing was that the text was one of the most rational texts available on the subject of rhetoric and it still didn't seem right. Moreover he had access to the authors, who were members of the department. He had asked and listened and talked and agreed with their answers in a rational way, but somehow still wasn't satisfied with them.

The text started with the premise that if rhetoric is to be taught at all at a University level it should be taught as a branch of reason, not as a mystic art. Therefore it emphasized a mastery of the rational foundations of communication in order to understand rhetoric. Elementary logic was introduced, elementary stimulus-response theory was brought in, and from these a progression was made to an understanding of how to develop an essay.

For the first year of teaching Phaedrus had been fairly content with this framework. He felt there was something wrong with it, but that the wrongness was not in this application of reason to rhetoric. The wrongness was in the old ghost of his dreams—rationality itself. He recognized it as the same wrongness that had been troubling him for years, and for which he had no solutions. He just felt that no writer ever learned to write by this squarish, by-the-numbers, objective, methodical approach. Yet that was all rationality offered and there was nothing to do about it without being irrational. And if there was one thing he had a clear mandate to do in this Church of Reason it was to be rational, so he had to let it go at that.

A few days later when Sarah trotted by again she stopped and said, "I'm so *happy* you're teaching *Quality* this quarter. Hardly anybody *is* these days."

"Well, *I* am," he said. "I'm definitely making a point of it."

"Good!" she said, and trotted on.

He returned to his notes but it wasn't long before thought about them was interrupted by a recall of her strange remark. What the hell was she talking about? *Quality?* Of *course* he was teaching Quality. Who wasn't? He continued with the notes.

Another thing that depressed him was prescriptive rhetoric, which supposedly had been done away with but was still around. This was the old slap-on-the-fingers-if-your-modifiers-were-caught-dangling stuff. *Correct* spelling, *correct* punctuation, *correct* grammar. Hundreds of itsy-bitsy rules for itsy-bitsy people. No one could remember all that stuff and concentrate on what he was trying to write about. It was all table manners, not derived from any sense of kindness or decency or humanity, but originally from an egotistic desire to look like gentlemen and ladies. Gentlemen and ladies had good table manners and spoke and wrote grammatically. It was what identified one with the upper classes.

In Montana, however, it didn't have this effect at all. It identified one, instead, as a stuck-up Eastern ass.

There was a minimum prescriptive-rhetoric requirement in the department, but like the other teachers he
scrupulously avoided any defense of prescriptive rhetoric rather than as a "requirement of the college."

Soon the thought interrupted again. *Quality?* There
was something irritating, even angering about that
question? He thought about it, and then thought some
more, and then looked out the window, and then
thought about it some more. *Quality?*

Four hours later he still sat there with his feet on the
window ledge and stared out into what had become a
dark sky. The phone rang, and it was his wife calling
him to find out what had happened. He told her he
would be home soon, but then forgot about this and
everything else. It wasn't until three o'clock in the
morning that he wearily confessed to himself that he
didn't have a clue as to what Quality was, picked up
his briefcase and headed home.

Most people would have forgotten about Quality at
this point, or just left it hanging suspended because
they were getting nowhere and had other things to do.
But he was so despondent about his own inability to
teach what he believed, he really didn't give a damn
about whatever else it was he was supposed to do, and
when he woke up the next morning there was Quality
staring him in the face. Three hours of sleep and he
was so tired he knew he wouldn't be up to giving a lecture that day, and besides, his notes had never been
completed, so he wrote on the blackboard: "Write a
350-word essay answering the question, What is *quality* in thought and statement?" Then he sat by the radiator while they wrote and thought about quality
himself.

At the end of the hour no one seemed to have finished, so he allowed the students to take their papers home. This class didn't meet again for two days, and that gave him some time to think about the question some more too. During that interim he saw some of the students walking between classes, nodded to them and got looks of anger and fear in return. He guessed they were having the same trouble he was.

Quality . . . you know what it is, yet you don't know what it is. But that's self-contradictory. But some things *are* better than others, that is, they have more quality. But when you try to say what the quality is, apart from the things that have it, it all goes *poof!* There's nothing to talk about. But if you can't say what Quality is, how do you know what it is, or how do you know that it even exists? If no one knows what it is, then for all practical purposes it doesn't exist at all. But for all practical purposes it really *does* exist. What else are the grades based on? Why else would people pay fortunes for some things and throw others in the trash pile? Obviously some things are better than others . . . but what's the "betterness"? . . . So round and round you go, spinning mental wheels and nowhere finding anyplace to get traction. What the hell is Quality? What *is* it?

PART III

16

Chris and I have had a good nights sleep and this morning have packed the backpacks carefully, and now have been going up the mountainside for about an hour. The forest here at the bottom of the canyon is mostly pine, with a few aspen and broad-leafed shrubs. Steep canyon walls rise way above us on both sides. Occasionally the trail opens into a patch of sunlight and grass that edges the canyon stream, but soon it reenters the deep shade of the pines. The earth of the trail is covered with a soft springy duff of pine needles. It is very quiet here.

Mountains like these and travelers in the mountains and events that happen to them here are found not only in Zen literature but in the tales of every major religion. The allegory of a physical mountain for the spiritual one that stands between each soul and its goal is an easy and natural one to make. Like those in the valley behind us, most people stand in sight of the spiritual mountains all their lives and never enter them, being content to listen to others who have been there and thus avoid the hardships. Some travel into the mountains accompanied by experienced guides who know the best and least dangerous routes by which they arrive at their destination. Still others, inexperienced and untrusting, attempt to make their own routes. Few of these are successful, but occasionally some, by sheer will and luck and grace, do make it. Once there they become more aware than any of the others that there's no single or fixed number of routes. There are as many routes as there are individual souls.

I want to talk now about Phaedrus' exploration into the meaning of the term *Quality,* an exploration which he saw as a route through the mountains of the spirit. As best I can puzzle it out, there were two distinct phases.

In the first phase he made no attempt at a rigid, systematic definition of what he was talking about. This was a happy, fulfilling and creative phase. It lasted most of the time he taught at the school back in the valley behind us.

The second phase emerged as a result of normal intellectual criticism of his lack of definition of what he was talking about. In this phase he made systematic, rigid statements about what Quality is, and worked out an enormous hierarchic structure of thought to support them. He literally had to move heaven and earth to arrive at this systematic understanding and when he was done felt he'd achieved an explanation of existence and our consciousness of it better than any that had existed before.

If it was truly a new route over the mountain it's certainly a needed one. For more than three centuries now the old routes common in this hemisphere have been undercut and almost washed out by the natural erosion and change of the shape of the mountain wrought by scientific truth. The early climbers established paths that were on firm ground with an accessibility that appealed to all, but today the Western routes are all but closed because of dogmatic inflexibility in the face of change. To doubt the literal meaning of the words of Jesus or Moses incurs hostility from most people, but it's just a fact that if Jesus or Moses were to appear today, unidentified, with the same message he spoke many years ago, his mental stability would be challenged. This isn't because what Jesus or Moses said was untrue or because modern society is in error but simply because the route they chose to reveal to others has lost relevance and comprehensibility. "Heaven above" fades from meaning when space-age consciousness asks, Where is "above"? But the fact that the old routes have tended, because of language rigidity, to lose their everyday meaning and become almost closed doesn't mean that the mountain is no longer there. It's there and will be there as long as consciousness exists.

Phaedrus' second metaphysical phase was a total disaster. Before the electrodes were attached to his head he'd lost everything tangible: money, property, children; even his rights as a citizen had been taken away from him by order of the court. All he had left was his one crazy lone dream of Quality, a map of a route across the mountain, for which he had sacrificed everything. Then, after the electrodes were attached, he lost that.

I will never know all that was in his head at that

time, nor will anyone else. What's left now is just fragments: debris, scattered notes, which can be pieced together but which leave huge areas unexplained.

When I first discovered this debris I felt like some agricultural peasant near the outskirts of, say, Athens, who occasionally and without much surprise plows up stones that have strange designs on them. I knew that these were part of some larger overall design that had existed in the past, but it was far beyond my comprehension. At first I deliberately avoided them, paid no attention to them because I knew these stones had caused some kind of trouble I should avoid. But I could see even then that they were a part of a huge structure of thought and I was curious about it in a secret sort of way.

Later, when I developed more confidence in my immunity to his affliction, I became interested in this debris in a more positive way and began to jot down the fragments amorphically, that is, without regard to form, in the order in which they occurred to me. Many of these amorphic statements have been supplied by friends. There are thousands of them now, and although only a small portion of them can fit into this Chautauqua, this Chautauqua is clearly based on them.

It is probably a long way from what he thought. When trying to re-create a whole pattern by deduction from fragments I am bound to commit errors and put down inconsistencies, for which I must ask some indulgence. In many cases the fragments are ambiguous; a number of different conclusions could be drawn. If something is wrong there's a good chance that the error isn't in what he thought but in my reconstruction of it, and a better reconstruction can later be found.

*   *   *

A whirr sounds and a partridge disappears through the trees.

"Did you see it?" says Chris.

"Yes," I say back.

"What was it?"

"A partridge."

"How do you know?"

"They rock back and forth like that when they fly," I say. I'm not sure of this but it sounds right. "They stay close to the ground, too."

"Oh," says Chris and we continue hiking. The rays of the sun create a cathedral effect through the pines.

Today now I want to take up the first phase of his journey into Quality, the nonmetaphysical phase, and this will be pleasant. It's nice to start journeys pleasantly, even when you know they won't end that way. Using his class notes as reference material I want to reconstruct the way in which Quality became a working concept for him in the teaching of rhetoric. His second phase, the metaphysical one, was tenuous and speculative, but this first phase, in which he simply taught rhetoric, was by all accounts solid and pragmatic and probably deserves to be judged on its own merits, independently of the second phase.

He'd been innovating extensively. He'd been having trouble with students who had nothing to say. At first he thought it was laziness but later it became apparent that it wasn't. They just couldn't think of anything to say.

One of them, a girl with strong-lensed glasses, wanted to write a five-hundred-word essay about the

United States. He was used to the sinking feeling that comes from statements like this, and suggested without disparagement that she narrow it down to just Bozeman.

When the paper came due she didn't have it and was quite upset. She had tried and tried but she just couldn't think of anything to say.

He had already discussed her with her previous instructors and they'd confirmed his impressions of her. She was very serious, disciplined and hardworking, but extremely dull. Not a spark of creativity in her anywhere. Her eyes, behind the thick-lensed glasses, were the eyes of a drudge. She wasn't bluffing him, she really couldn't think of anything to say, and was upset by her inability to do as she was told.

It just stumped him. Now *he* couldn't think of anything to say. A silence occurred, and then a peculiar answer: "Narrow it down to the *main street* of Bozeman." It was a stroke of insight.

She nodded dutifully and went out. But just before her next class she came back in *real* distress, tears this time, distress that had obviously been there for a long time. She still couldn't think of anything to say, and couldn't understand why, if she couldn't think of anything about *all* of Bozeman, she should be able to think of something about just one street.

He was furious. "You're not *looking*!" he said. A memory came back of his own dismissal from the University for having *too much* to say. For every fact there is an *infinity* of hypotheses. The more you *look* the more you *see*. She really wasn't looking and yet somehow didn't understand this.

He told her angrily, "Narrow it down to the *front* of

one building on the main street of Bozeman. The Opera House. Start with the upper left-hand brick."

Her eyes, behind the thick-lensed glasses, opened wide.

She came in the next class with a puzzled look and handed him a five-thousand-word essay on the front of the Opera House on the main street of Bozeman, Montana. "I sat in the hamburger stand across the street," she said, "and started writing about the first brick, and the second brick, and then by the third brick it all started to come and I couldn't stop. They thought I was crazy, and they kept kidding me, but here it all is. I don't understand it."

Neither did he, but on long walks through the streets of town he thought about it and concluded she was evidently stopped with the same kind of blockage that had paralyzed him on his first day of teaching. She was blocked because she was trying to repeat, in her writing, things she had already heard, just as on the first day he had tried to repeat things he had already decided to say. She couldn't think of anything to write about Bozeman because she couldn't recall anything she had heard worth repeating. She was strangely unaware that she could look and see freshly for herself, as she wrote, without primary regard for what had been said before. The narrowing down to one brick destroyed the blockage because it was so obvious she *had* to do some original and direct seeing.

He experimented further. In one class he had everyone write all hour about the back of his thumb. Everyone gave him funny looks at the beginning of the hour, but everyone did it, and there wasn't a single complaint about "nothing to say."

In another class he changed the subject from the thumb to a coin, and got a full hour's writing from every student. In other classes it was the same. Some asked, "Do you have to write about both sides?" Once they got into the idea of seeing directly for themselves they also saw there was no limit to the amount they could say. It was a confidence-building assignment too, because what they wrote, even though seemingly trivial, was nevertheless their own thing, not a mimicking of someone else's. Classes where he used that coin exercise were always less balky and more interested.

As a result of his experiments he concluded that imitation was a real evil that had to be broken before real rhetoric teaching could begin. This imitation seemed to be an external compulsion. Little children didn't have it. It seemed to come later on, possibly as a result of school itself.

That sounded right, and the more he thought about it the more right it sounded. Schools teach you to imitate. If you don't imitate what the teacher wants you get a bad grade. Here, in college, it was more sophisticated, of course; you were supposed to imitate the teacher in such a way as to convince the teacher you were not imitating, but taking the essence of the instruction and going ahead with it on your own. That got you A's. Originality on the other hand could get you anything—from A to F. The whole grading system cautioned against it.

He discussed this with a professor of psychology who lived next door to him, an extremely imaginative teacher, who said, "Right. Eliminate the whole degree-and-grading system and then you'll get real education."

Phaedrus thought about this, and when weeks later a very bright student couldn't think of a subject for a term paper, it was still on his mind, so he gave it to her as a topic. She didn't like the topic at first, but agreed to take it anyway.

Within a week she was talking about it to everyone, and within two weeks had worked up a superb paper. The class she delivered it to didn't have the advantage of two weeks to think about the subject, however, and was quite hostile to the whole idea of eliminating grades and degrees. This didn't slow her down at all. Her tone took on an old-time religious fervor. She begged the other students to *listen,* to understand this was really *right.* "I'm not saying this for *him,*" she said and glanced at Phaedrus. "It's for *you.*"

Her pleading tone, her religious fervor, greatly impressed him, along with the fact that her college entrance examinations had placed her in the upper one percent of the class. During the next quarter, when teaching "persuasive writing," he chose this topic as a "demonstrator," a piece of persuasive writing he worked up by himself, day by day, in front of and with the help of the class.

He used the demonstrator to avoid talking in terms of principles of composition, all of which he had deep doubts about. He felt that by exposing classes to his own sentences as he made them, with all the misgivings and hang-ups and erasures, he would give a more honest picture of what writing was like than by spending class time picking nits in completed student work or holding up the completed work of masters for emulation. This time he developed the argument that the whole grading system and degree should be eliminated, and to make it something that truly involved the students in what they were hearing, he withheld all grades during the quarter.

Just up above the top of the ridge the snow can be seen now. On foot it's many days away though. The rocks below it are too steep for a direct hiking climb, particularly with the heavy loads we are carrying, and Chris is way too young for any kind of ropes-and-pitons stuff. We must cross over the forested ridge we are now approaching, enter another canyon, follow it to its end and then come back at an upward angle along to the ridge. Three days hard to the snow. Four days easy. If we don't show up in nine, DeWeese will start looking for us.

We stop for a rest, sit down and brace against a tree so that we don't topple over backward from the packs. After a while I reach around over my shoulder, take the machete from the top of my pack and hand it to Chris.

"See those two aspens over there? The straight ones? At the edge?" I point to them. "Cut those down about a foot from the ground."

"Why?"

"We'll need them later for hiking sticks and tent poles."

Chris takes the machete, starts to rise but then settles back again. "You cut them," he says.

So I take the machete and go over and cut the poles. They both cut neatly in one swing, except for the final strip of bark, which I sever with the back hook of the machete. Up in the rocks you need the poles for balancing and the pine up above is no good for poles, and this is about the last of the aspen here. It bothers me a little though that Chris is turning down work. Not a good sign in the mountains.

A short rest and then on we go. It'll take a while to get used to this load. There's a negative reaction to all the weight. As we go on though, it'll become more natural. . . .

Phaedrus' argument for the abolition of the degree-and-grading system produced a nonplussed or negative reaction in all but a few students at first, since it seemed, on first judgment, to destroy the whole University system. One student laid it wide open when she said with complete candor, "Of course you can't eliminate the degree and grading system. After all, that's what we're here for."

She spoke the complete truth. The idea that the majority of students attend a university for an education independent of the degree and grades is a little hypocrisy everyone is happier not to expose. Occasionally some students do arrive for an education but rote and the mechanical nature of the institution soon converts them to a less idealistic attitude.

The demonstrator was an argument that elimination of grades and degrees would destroy this hypocrisy. Rather than deal with generalities it dealt with the specific career of an imaginary student who more or less typified what was found in the classroom, a student completely conditioned to work for a grade rather than for the knowledge the grade was supposed to represent.

Such a student, the demonstrator hypothesized, would go to his first class, get his first assignment and probably do it out of habit. He might go to his second and third as well. But eventually the novelty of the course would wear off and, because his academic life was not his only life, the pressure of other obligations or desires would create circumstances where he just would not be able to get an assignment in.

Since there was no degree or grading system he would incur no penalty for this. Subsequent lectures which presumed he'd completed the assignment might be a little more difficult to understand, however, and this difficulty, in turn, might weaken his interest to a point where the next assignment, which he would find quite hard, would also be dropped. Again no penalty.

In time his weaker and weaker understanding of what the lectures were about would make it more and more difficult for him to pay attention in class. Eventually he would see he wasn't learning much; and facing the continual pressure of outside obligations, he would stop studying, feeling guilty about this and stop attending class. Again, no penalty would be attached.

But what had happened? The student, with no hard feelings on anybody's part, would have flunked himself out. Good! This is what should have happened. He wasn't there for a real education in the first place and had no real business there at all. A large amount of money and effort had been saved and there would be no stigma of failure and ruin to haunt him the rest of his life. No bridges had been burned.

The student's biggest problem was a slave mentality which had been built into him by years of carrot-and-whip grading, a mule mentality which said, "If you don't whip me, I won't work." He didn't get whipped. He didn't work. And the cart of civilization, which he supposedly was being trained to pull, was just going to have to creak along a little slower without him.

This is a tragedy, however, only if you presume that the cart of civilization, "the system," is pulled by mules. This is a common, vocational, "location" point of view, but it's not the Church attitude.

The Church attitude is that civilization, or "the system" or "society" of whatever you want to call it, is best served not by mules but by free men. The purpose of abolishing grades and degrees is not to punish mules or to get rid of them but to provide an environment in which that mule can turn into a free man.

The hypothetical student, still a mule, would drift around for a while. He would get another kind of education quite as valuable as the one he'd abandoned, in what used to be called the "school of hard knocks." Instead of wasting money and time as a high-status mule, he would now have to get a job as a low-status mule, maybe as a mechanic. Actually his *real* status would go up. He would be making a contribution for a change. Maybe that's what he would do for the rest of his life. Maybe he'd found his level. But don't count on it.

In time—six months; five years, perhaps—a change could easily begin to take place. He would become less and less satisfied with a kind of dumb, day-to-day shopwork. His creative intelligence, stifled by too much theory and too many grades in college, would now become reawakened by the boredom of the shop. Thousands of hours of frustrating mechanical problems would have made him more interested in machine design. He would like to design machinery himself.

He'd think he could do a better job. He would try modifying a few engines, meet with success, look for more success, but feel blocked because he didn't have the theoretical information. He would discover that when before he felt stupid because of his lack of interest in theoretical information, he'd now find a brand of theoretical information which he'd have a lot of respect for, namely, mechanical engineering.

So he would come back to our degreeless and gradeless school, but with a difference. He'd no longer be a grade-motivated person. He'd be a knowledge-motivated person. He would need no external pushing to learn. His push would come from inside. He'd be a free man. He wouldn't need a lot of discipline to shape him up. In fact, if the instructors assigned him were slacking on the job he would be likely to shape *them* up by asking rude questions. He'd be there to learn something, would be paying to learn something and they'd better come up with it.

Motivation of this sort, once it catches hold, is a ferocious force, and in the gradeless, degreeless institution where our student would find himself, he wouldn't stop with rote engineering information. Physics and mathematics were going to come within his sphere of interest because he'd see he needed them. Metallurgy and electrical engineering would come up for attention. And, in the process of intellectual maturing that these abstract studies gave him, he would be likely to branch out into other theoretical areas that weren't directly related to machines but had become a part of a newer larger goal. This larger goal wouldn't be the imitation of education in Universities today, glossed over and concealed by grades and degrees that give the appearance of something happening when, in fact, almost nothing is going on. It would be the real thing.

Such was Phaedrus' demonstrator, his unpopular argument, and he worked on it all quarter long, building it up and modifying it, arguing for it, defending it. All quarter long papers would go back to the students with comments but no grades, although the grades were entered in a book.

As I said before, at first almost everyone was sort of nonplussed. The majority probably figured they were stuck with some idealist who thought removal of grades would make them happier and thus work harder, when it was obvious that without grades everyone would just loaf. Many of the students with A records in previous quarters were contemptuous and angry at first, but because of their acquired self-discipline went ahead and did the work anyway. The B students and high-C students missed some of the early assignments or turned in sloppy work. Many of the low-C and D students didn't even show up for class. At this time another teacher asked him what he was going to do about this lack of response.

"Outwait them," he said.

His lack of harshness puzzled the students at first, then made them suspicious. Some began to ask sarcastic questions. These received soft answers and the lectures and speeches proceeded as usual, except with no grades.

Then a hoped-for phenomenon began. During the third or fourth week some of the A students began to get nervous and started to turn in superb work and hang around after class with questions that fished for some indication as to how they were doing. The B and high-C students began to notice this and work a little and bring up the quality of their papers to a more usual level. The low C, D and future F's began to show up for class just to see what was going on.

After midquarter an even more hoped-for phenomenon took place. The A-rated students lost their nervousness and became active participants in everything that went on with a friendliness that was uncommon in a grade-getting class. At this point the B and C students were in a panic, and turned in stuff that looked as though they'd spent hours of painstaking work on it. The D's and F's turned in satisfactory assignments.

In the final weeks of the quarter, a time when normally everyone knows what his grade will be and just sits back half asleep, Phaedrus was getting a kind of class participation that made other teachers take notice. The B's and C's had joined the A's in friendly free-for-all discussions that made the class seem like a successful party. Only the D's and F's sat frozen in their chairs, in a complete internal panic.

The phenomenon of relaxation and friendliness was explained later by a couple of students who told him, "A lot of us got together outside of class to try to figure out how to beat this system. Everyone decided the best way was just to figure you were going to fail and then go ahead and do what you could anyway. Then you start to relax. Otherwise you go out of your mind!"

The students added that once you got used to it it wasn't so bad, you were more interested in the subject matter, but repeated that it wasn't easy to get used to.

At the end of the quarter the students were asked to write an essay evaluating the system. None of them knew at the time of writing what his or her grade would be. Fifty-four percent opposed it. Thirty-seven percent favored it. Nine percent were neutral.

On the basis of one man, one vote, the system was very unpopular. The majority of students definitely wanted their grades as they went along. But when Phaedrus broke down the returns according to the grades that were in his book—and the grades were not out of line with grades predicted by previous classes and entrance evaluations—another story was told. The A students were 2 to 1 in favor of the system. The B and C students were evenly divided. And the D's and F's were *unanimously* opposed!

This surprising result supported a hunch he had had for a long time: that the brighter, more serious students were the *least* desirous of grades, possibly because they were more interested in the subject matter of the course, whereas the dull or lazy students were the *most* desirous of grades, possibly because grades told them if they were getting by.

As DeWeese said, from here straight south you can go seventy-five miles through nothing but forests and snow without ever encountering a road, although there are roads to the east and the west. I've arranged it so that if things work out badly at the end of the second day we'll be near a road that can get us back fast. Chris doesn't know about this, and it would hurt his YMCA-camp sense of adventure to tell him, but after enough trips into the high country, the YMCA desire for adventure diminishes and the more substantial benefits of cutting down risks appear. This country can be dangerous. You take one bad step in a million, sprain an ankle, and then you find out how far from civilization you really are.

This is apparently a seldom-entered canyon this far up. After another hour of hiking we see that the trail is about gone.

Phaedrus thought withholding grades was good, according to his notes, but he didn't give it scientific value. In a true experiment you keep constant every cause you can think of except one, and then see what the effects are of varying that one cause. In the classroom you can never do this. Student knowledge, student attitude, teacher attitude, all change from all kinds of causes which are uncontrollable and mostly unknowable. Also, the observer in this case is himself one of the causes and can never judge his effects without altering his effects. So he didn't attempt to draw any hard conclusions from all this, he just went ahead and did what he liked.

The movement from this to his inquiry into Quality took place because of a sinister aspect of grading that the withholding of grades exposed. Grades really cover up failure to teach. A bad instructor can go through an entire quarter leaving absolutely nothing memorable in the minds of his class, curve out the scores on an irrelevant test, and leave the impression that some have learned and some have not. But if the grades were removed the class is forced to wonder each day what it's *really* learning. The questions, What's being taught? What's the goal? How do the lectures and assignments accomplish the goal become ominous. The removal of grades exposes a huge and frightening vacuum.

What was Phaedrus trying to do, anyway? This question became more and more imperative as he went on. The answer that had seemed right when he started now made less and less sense. He had wanted his students to become creative by deciding for themselves what was good writing instead of asking him all the time. The real purpose of withholding the grades was to force them to look within themselves, the only place they would ever get a really right answer.

But now this made no sense. If they already knew what was good and bad, there was no reason for them to take the course in the first place. The fact that they were there as students presumed they did *not* know what was good or bad. That was his job as instructor— to tell them what was good or bad. The whole idea of individual creativity and expression in the classroom was really basically opposed to the whole idea of the University.

For many of the students, this withholding created a Kafkaesque situation in which they saw they were to be punished for failure to do something but no one would tell them what they were supposed to do. They looked within themselves and saw nothing and looked at Phaedrus and saw nothing and just sat there helpless, not knowing what to do. The vacuum was deadly. One girl suffered a nervous breakdown. You cannot withhold grades and sit there and create a goalless vacuum. You have to provide some goal for a class to work toward that will fill that vacuum. This he wasn't doing.

He couldn't. He could think of no possible way he could tell them what they should work toward without falling back into the trap of authoritarian, didactic teaching. But how can you put on the blackboard the mysterious internal goal of each creative person?

The next quarter he dropped the whole idea and went back to regular grading, discouraged, confused, feeling he was right but somehow it had come out all wrong. When spontaneity and individuality and really good original stuff occurred in a classroom it was in spite of the instruction, not because of it. This seemed to make sense. He was ready to resign. Teaching dull conformity to hateful students wasn't what he wanted to do.

He'd heard that Reed College in Oregon withheld grades until graduation, and during the summer vacation he went there but was told the faculty was divided on the value of withholding grades and that no one was tremendously happy about the system. During the rest of the summer his mood became depressed and lazy. He and his wife camped a lot in those mountains. She asked why he was so silent all the time but he couldn't say why. He was just stopped. Waiting. For that missing seed crystal of thought that would suddenly solidify everything.

17

It's looking bad for Chris. For a while he was way ahead of me and now he sits under a tree and rests. He doesn't look at me, and that's how I know it's bad.

I sit down next to him and his expression is distant. His face is flushed and I can see he's exhausted. We sit and listen to the wind through the pines.

I know eventually he'll get up and keep going but *he* doesn't know this, and is afraid to face the possibility that his fear creates: that he may not be able to climb the mountain at all. I remember something Phaedrus had written about these mountains and tell it to Chris now.

"Years ago," I tell him, "your mother and I were at the timberline not so far from here and we camped near a lake with a marsh on one side."

He doesn't look up but he's listening.

"At about dawn we heard falling rocks and we thought it must be an animal, except that animals don't usually clatter around. Then I heard a squishing sound in the marsh and we were really wide-awake. I got out of the sleeping bag slowly and got our revolver from my jacket and crouched by a tree."

Now Chris's attention is distracted from his own problems.

"There was another squish," I say. "I thought it could be horses with dudes packing in, but not at this hour. Another *squish*! And a loud *galoomph*! That's no horse! And a *Galoomph*! and a GALOOMPH! And there, in the dim grey light of dawn coming straight for me through the muck of the marsh, was the biggest bull moose I ever saw. Horns as wide as a man is tall. Next to the grizzly the most dangerous animal in the mountains. Some say the worst."

Chris's eyes are bright again.

"*GALOOMPH!* I cocked the hammer on the revolver, thinking a thirty-eight Special wasn't very much for a moose. *GALOOMPH!* He didn't SEE me! *GALOOMPH!* I couldn't get out of his way. Your mother was in the sleeping bag right in his path. *GALOOMPH!* What a GIANT! *GALOOMPH!* He's *ten yards* away! *GALOOMPH!* I stand up and take aim. *GALOOMPH! . . . GALOOMPH! . . . GALOOMPH!* . . . He stops, THREE YARDS AWAY, and sees me. . . . The gunsights lie right between his eyes. . . . We're motionless."

I reach around into my pack and get out some cheese.

"Then what happened?" Chris asks.

"Wait until I cut off some of this cheese."

I remove my hunting knife and hold the cheese wrapper so that my fingers don't get on it. I slice out a quarter-inch hunk and hold it out for him.

He takes it. "Then what happened?"

I watch until he takes his first bite. "That bull moose looked at me for what must have been five seconds. Then he looked down at your mother. Then he looked at me again, and at the revolver which was practically lying on top of his big round nose. And then he smiled and slowly walked away."

"Oh," says Chris. He looks disappointed.

"Normally when they're confronted like that they'll charge," I say, "but he just thought it was a nice morning, and we were there first, so why make trouble? And that's why he smiled."

"Can they *smile*?"

"No, but it looked that way."

I put the cheese away and add, "Later on that day we were jumping from boulder to boulder down the side of a slope. I was about to land on a great big brown boulder when all of a sudden the great big brown boulder jumped into the air and ran off into the woods. It was the same moose. . . . I think that moose must have been pretty sick of us that day."

I help Chris get to his feet. "You were going a little too fast," I say. "Now the mountainside's becoming steep and we have to go slowly. If you go too fast you get winded and when you get winded you get dizzy and that weakens your spirit and you think, I can't do it. So go slow for a while."

"I'll stay behind you," he says.

"Okay."

We walk now away from the stream we were following, up the canyon side at the shallowest angle I can find.

Mountains should be climbed with as little effort as

possible and without desire. The reality of your own nature should determine the speed. If you become restless, speed up. If you become winded, slow down. You climb the mountain in an equilibrium between restlessness and exhaustion. Then, when you're no longer thinking ahead, each footstep isn't just a means to an end but a unique event in itself. *This* leaf has jagged edges. *This* rock looks loose. From *this* place the snow is less visible, even though closer. These are things you should notice anyway. To live only for some future goal is shallow. It's the sides of the mountain which sustain life, not the top. Here's where things grow.

But of course, without the top you can't have any sides. It's the top that *defines* the sides. So on we go . . . we have a long way . . . no hurry . . . just one step after the next . . . with a little Chautauqua for entertainment. . . . Mental reflection is so much more interesting than TV it's a shame more people don't switch over to it. They probably think what they hear is unimportant but it never is.

There's a large fragment concerning Phaedrus' first class after he gave that assignment on "What is quality in thought and statement?" The atmosphere was explosive. Almost everyone seemed as frustrated and angered as he had been by the question.

"How are *we* supposed to know what quality is?" they said. "You're supposed to tell *us*!"

Then he told them he couldn't figure it out either and really wanted to know. He had assigned it in the hope that somebody would come up with a good answer.

That ignited it. A roar of indignation shook the room. Before the commotion had settled down another teacher had stuck his head in the door to see what the trouble was.

"It's all right," Phaedrus said. "We just accidentally stumbled over a genuine question, and the shock is hard to recover from." Some students looked curious at this, and the noise simmered down.

He then used the occasion for a short return to his theme of "Corruption and Decay in the Church of Reason." It was a measure of this corruption, he said, that students should be outraged by someone trying to *use* them to seek the truth. You were supposed to *fake* this search for the truth, to *imitate* it. To actually *search* for it was a damned imposition.

The truth was, he said, that he genuinely did want to know what they thought, not so that he could put a grade on it, but because he really wanted to know.

They looked puzzled.

"I sat there all night long," one said.

"I was ready to cry, I was so mad," a girl next to the window said.

"You should warn us," a third said.

"How could I warn you," he said, "when I had no idea how you'd react?"

Some of the puzzled ones looked at him with a first dawning. He wasn't playing games. He really wanted to know.

A most peculiar person.

Then someone said, "What do *you* think?"

"I don't *know,*" he answered.

"But what do you *think?*"

He paused for a long time. "I think there is such a

thing as Quality, but that as soon as you try to define it, something goes haywire. You can't do it."

Murmurs of agreement.

He continued, "Why this is, I don't know. I thought maybe I'd get some ideas from your papers. I just don't know."

This time the class was silent.

In subsequent classes that day there was some of the same commotion, but a number of students in each class volunteered friendly answers that told him the class had been discussed during lunch.

A few days later he worked up a definition of his own and put it on the blackboard to be copied for posterity. The definition was: "Quality is a characteristic of thought and statement that is recognized by a non-thinking process. Because definitions are a product of rigid, formal thinking, quality cannot be defined."

The fact that this "definition" was actually a refusal to define did not draw comment. The students had no formal training that would have told them his statement was, in a formal sense, completely irrational. If you can't define something you have no formal rational way of knowing that it exists. Neither can you *really* tell anyone else what it is. There is, in fact, no formal difference between inability to define and stupidity. When I say, "Quality cannot be defined," I'm really saying formally, "I'm stupid about Quality."

Fortunately the students didn't know this. If they'd come up with these objections he wouldn't have been able to answer them at the time.

But then, below the definition on the blackboard, he wrote, "But even though Quality cannot be defined, *you know what Quality is!*" and the storm started all over again.

"Oh, no, we don't!"

"Oh, yes, you do."

"Oh, *no,* we *don't!*"

"Oh, *yes,* you *do!*" he said and he had some material ready to demonstrate it to them.

He had selected two examples of student composition. The first was a rambling, disconnected thing with interesting ideas that never built into anything. The second was a magnificent piece by a student who was mystified himself about why it had come out so well. Phaedrus read both, then asked for a show of hands on who thought the first was best. Two hands went up. He asked how many liked the second better. Twenty-eight hands went up.

"Whatever it is," he said, "that caused the overwhelming majority to raise their hands for the second one is what I mean by Quality. So *you* know what it is."

There was a long reflective silence after this, and he just let it last.

This was just intellectually outrageous, and he knew it. He wasn't teaching anymore, he was indoctrinating. He had erected an imaginary entity, defined it as incapable of definition, told the students over their own protests that they knew what it was, and demonstrated this by a technique that was as confusing logically as the term itself. He was able to get away with this because logical refutation required more talent than any of the students had. In subsequent days he continually invited their refutations, but none came. He improvised further.

To reinforce the idea that they already knew what

Quality was he developed a routine in which he read four student papers in class and had everyone rank them in estimated order of Quality on a slip of paper. He did the same himself. He collected the slips, tallied them on the blackboard and averaged the rankings for an overall class opinion. Then he would reveal his own rankings, and this would almost always be close to, if not identical with, the class average. Where there were differences it was usually because two papers were close in quality.

At first the classes were excited by this exercise, but as time went on they became bored. What he meant by Quality was obvious. They obviously knew what it was too, and so they lost interest in listening. Their question now was "All right, we know what Quality is. How do we get it?"

Now, at last, the standard rhetoric texts came into their own. The principles expounded in them were no longer rules to rebel against, not ultimates in themselves, but just techniques, gimmicks, for producing what really counted and stood independently of the techniques—Quality. What had started out as a heresy from traditional rhetoric turned into a beautiful introduction to it.

He singled out aspects of Quality such as unity, vividness, authority, economy, sensitivity, clarity, emphasis, flow, suspense, brilliance, precision, proportion, depth and so on; kept each of these as poorly defined as Quality itself, but demonstrated them by the same class reading techniques. He showed how the aspect of Quality called unity, the hanging-togetherness of a story, could be improved with a technique called an outline. The authority of an argument could be jacked up with a technique called footnotes, which gives authoritative reference. Outlines and footnotes are standard things taught in all freshman composition classes, but now as devices for improving Quality they had a purpose. And if a student turned in a bunch of dumb references or a sloppy outline that showed he was just fulfilling an assignment by rote, he could be told that while his paper may have fulfilled the letter of the assignment it obviously didn't fulfill the goal of Quality, and was therefore worthless.

Now, in answer to that eternal student question, How do I *do* this? that had frustrated him to the point of resignation, he could reply, "It doesn't make a bit of difference *how* you do it! Just so it's good." The reluctant student might ask in class, "But how do we know what's good?" but almost before the question was out of his mouth he would realize the answer had already been supplied. Some other student would usually tell him, "You just *see* it." If he said, "No, I don't," he'd be told, "Yes, you do. He proved it." The student was finally and completely trapped into making quality judgments for himself. And it was just exactly this and nothing else that taught him to write.

Up to now Phaedrus had been compelled by the academic system to say what he wanted, even though he knew that this forced students to conform to artificial forms that destroyed their own creativity. Students who went along with his rules were then condemned for their inability to be creative or produce a piece of work that reflected their own personal standards of what is good.

Now that was over with. By reversing a basic rule that all things which are to be taught must first be defined, he had found a way out of all this. He was pointing to no principle, no rule of good writing, no theory—but he was pointing to something, nevertheless, that was very real, whose reality they couldn't deny. The vacuum that had been created by the withholding of grades was suddenly filled with the positive goal of Quality, and the whole thing fit together. Students, astonished, came by his office and said, "I used to just *hate* English. Now I spend more time on it than anything else." Not just one or two. Many. The whole Quality concept was beautiful. It worked. It was that mysterious, individual, internal goal of each creative person, on the blackboard at last.

I turn to see how Chris is doing. His face looks tired.

I ask, "How do you feel?"

"Okay," he says, but his tone is defiant.

"We can stop anywhere and camp," I say.

He flashes a fierce look at me, and so I say nothing more. Soon I see he's working his way around me on the slope. With what must be great effort he pulls ahead. We go on.

Phaedrus got this far with his concept of Quality because he deliberately refused to look outside the immediate classroom experience. Cromwell's statement, "No one ever travels so high as he who knows not where he is going," applied at this point. He didn't know where he was going. All he knew was that it worked.

In time, however, he wondered *why* it worked, especially when he already knew it was irrational. Why should an irrational method work when rational methods were all so rotten? He had an intuitive feeling, growing rapidly, that what he had stumbled on was no small gimmick. It went far beyond. How far, he didn't know.

This was the beginning of the crystallization that I talked about before. Others wondered at the time, "Why should he get so excited about 'quality'?" But they saw only the word and its rhetoric context. They didn't see his past despair over abstract questions of existence itself that he had abandoned in defeat.

If anyone else had asked, What is Quality? it *would* have been just another question. But when *he* asked it, because of his past, it spread out for him like waves in all directions simultaneously, not in a hierarchic structure, but in a concentric one. At the center, generating the waves, was Quality. As these waves of thought expanded for him I'm sure he fully expected each wave to reach some shore of existing patterns of thought so that he had a kind of unified relationship with these thought structures. But the shore was never reached until the end, if it appeared at all. For him there was nothing but ever expanding waves of crystallization. I'll now try to follow these waves of crystallization, the second phase of his exploration into quality, as best I can.

Up ahead all of Chris's movements seem tired and angry. He stumbles on things, lets branches tear at him, instead of pulling them to one side.

I'm sorry to see this. Some blame can be put on the YMCA camp he attended for two weeks just before we started. From what he's told me, they made a big ego thing out of the whole outdoor experience. A proof-of-manhood thing. He began in a lowly class they were careful to point out was rather disgraceful to be in . . . original sin. Then he was allowed to prove himself with a long series of accomplishments—swimming, rope tying . . . he mentioned a dozen of them, but I've forgotten them.

It made the kids at camp much more enthusiastic and cooperative when they had ego goals to fulfill, I'm sure, but ultimately that kind of motivation is destructive. Any effort that has self-glorification as its final endpoint is bound to end in disaster. Now we're paying the price. When you try to climb a mountain to prove how big you are, you almost never make it. And even if you do it's a hollow victory. In order to sustain the victory you have to prove yourself again and again in some other way, and again and again and again, driven forever to fill a false image, haunted by the fear that the image is not true and someone will find out. That's never the way.

Phaedrus wrote a letter from India about a pilgrimage to holy Mount Kailas, the source of the Ganges and the abode of Shiva, high in the Himalayas, in the company of a holy man and his adherents.

He never reached the mountain. After the third day he gave up, exhausted, and the pilgrimage went on without him. He said he had the physical strength but that physical strength wasn't enough. He had the intellectual motivation but that wasn't enough either. He didn't think he had been arrogant but thought that he was undertaking the pilgrimage to broaden *his* experience, to gain understanding for *himself.* He was trying to use the mountain for his own purposes and the pilgrimage too. He regarded himself as the fixed entity, not the pilgrimage or the mountain, and thus wasn't ready for it. He speculated that the other pilgrims, the ones who reached the mountain, probably sensed the holiness of the mountain so intensely that each footstep was an act of devotion, an act of submission to this holiness. The holiness of the mountain infused into their own spirits enabled them to endure far more than anything he, with his greater physical strength, could take.

To the untrained eye ego-climbing and selfless climbing may appear identical. Both kinds of climbers place one foot in front of the other. Both breathe in and out at the same rate. Both stop when tired. Both go forward when rested. But what a difference! The ego-climber is like an instrument that's out of adjustment. He puts his foot down an instant too soon or too late. He's likely to miss a beautiful passage of sunlight through the trees. He goes on when the sloppiness of his step shows he's tired. He rests at odd times. He looks up the trail trying to see what's ahead even when he knows what's ahead because he just looked a second before. He goes too fast or too slow for the conditions and when he talks his talk is forever about somewhere else, something else. He's here but he's not here. He rejects the here, is unhappy with it, wants to be farther up the trail but when he gets there will be just as unhappy because then *it* will be "here." What he's looking for, what he wants, is all around him, but he doesn't want that because it *is* all around him. Every step's an effort, both physically and spiritually, because he imagines his goal to be external and distant.

That seems to be Chris's problem now.

18

There's an entire branch of philosophy concerned with the definition of Quality, known as esthetics. Its question, What is meant by *beautiful?* goes back to antiquity. But when he was a student of philosophy Phaedrus had recoiled violently from this entire branch of knowledge. He had almost deliberately failed the one course in it he had attended and had written a number of papers subjecting the instructor and materials to outrageous attack. He hated and reviled everything.

It wasn't any particular esthetician who produced this reaction in him. It was all of them. It wasn't any particular point of view that outraged him so much as the idea that Quality should be subordinated to *any* point of view. The intellectual process was forcing Quality into its servitude, prostituting it. I think that was the source of his anger.

He wrote in one paper, "These estheticians think their subject is some kind of peppermint bonbon they're entitled to smack their fat lips on; something to be devoured; something to be intellectually knifed, forked and spooned up bit by bit with appropriate delicate remarks and I'm ready to throw up. What they smack their lips on is the putrescence of something they long ago killed."

Now, as the first step of the crystallization process, he saw that when Quality is kept undefined by definition, the entire field called esthetics is wiped out . . . completely disenfranchised . . . kaput. By refusing to define Quality he had placed it entirely outside the analytic process. If you can't define Quality, there's no way you can subordinate it to any intellectual rule. The estheticians can have nothing more to say. Their whole field, definition of Quality, is gone.

The thought of this completely thrilled him. It was like discovering a cancer cure. No more explanations of what art is. No more wonderful critical schools of experts to determine rationally where each composer had succeeded or failed. All of them, every last one of those know-it-alls, would finally have to shut up. This was no longer just an interesting idea. This was a dream.

I don't think anyone really saw what he was up to at first. They saw an intellectual delivering a message that had all the trappings of a rational analysis of a teaching situation. They didn't see he had a purpose completely opposite to any they were used to. He wasn't furthering rational analysis. He was blocking it. He was turning the method of rationality against itself, turning it against his own kind, in defense of an irrational concept, an undefined entity called Quality.

He wrote: "(1) Every instructor of English composition knows what quality is. (Any instructor who does not should keep this fact carefully concealed, for this would certainly constitute proof of incompetence.) (2) Any instructor who thinks quality of writing can and should be defined before teaching it can and should go ahead and define it. (3) All those who feel that quality of writing does exist but cannot be defined, but that quality should be taught anyway, can benefit by the following method of teaching pure quality in writing without defining it."

He then went ahead and described some of the methods of comparison that had evolved in the classroom.

I think he really did hope that someone would come along, challenge him and try to define Quality for him. But no one ever did.

However, that little parenthetic statement about inability to define Quality as proof of incompetence did raise eyebrows within the department. He was, after all, the junior member, and not really expected to provide standards quite yet for his seniors' performance.

His right to say as he pleased was valued, and the senior members actually seemed to enjoy his independence of thought and support him in a churchlike way. But contrary to the belief of many opponents of academic freedom, the church attitude has never been that a teacher should be allowed to blather anything that comes into his head without any accountability at all. The church attitude is simply that the accountability must be to the God of Reason, not to the idols of political power. The fact that he was insulting people was irrelevant to the truth or falsehood of what he was saying and he couldn't ethically be struck down for this. But what they were prepared to strike him down for, ethically and with gusto, was any indication that he wasn't making sense. He could do anything he wanted as long as he justified it in terms of reason.

But how the hell do you ever justify, in terms of reason, a refusal to define something? Definitions are the *foundation* of reason. You can't reason without them. He could hold off the attack for a while with fancy dialectical footwork and insults about competence and incompetence, but sooner or later he had to come up with something more substantial than that. His attempt to come up with something substantial led to further crystallization beyond the traditional limits of rhetoric and into the domain of philosophy.

Chris turns and flashes a tormented look at me. It won't be long now. Even before we left there were clues this was coming. When DeWeese told a neighbor I was experienced in the mountains Chris showed a big flash of admiration. It was a large thing in his eyes. He should be done for soon, and then we can stop for the day.

Oop! There he goes. He's fallen down. He's not getting up. It was an awfully neat fall, not very accidental-looking. Now he looks at me with hurt and anger, searching for condemnation from me. I don't show him any. I sit down next to him and see he's almost defeated.

"Well," I say, "we can stop here, or we can go ahead, or we can go back. Which do you want to do?"

"I don't care," he says, "I don't want to . . ."

"You don't want to what?"

"I don't care!" he says, angrily.

"Then since you don't care, we'll keep on going," I say, trapping him.

"I don't like this trip," he says. "It isn't any fun. I thought it was going to be fun."

Some anger catches me off guard too. "That may be true," I reply, "but it's a hell of a thing to say."

I see a sudden flick of fear in his eyes as he gets up.

We go on.

The sky over the other wall of the canyon has become overcast, and the wind in the pines around us has become cool and ominous.

At least the coolness makes it easier hiking. . . .

I was talking about the first wave of crystallization outside of rhetoric that resulted from Phaedrus' refusal to define Quality. He had to answer the question, If you can't define it, what makes you think it exists?

His answer was an old one belonging to a philosophic school that called itself *realism*. "A thing exists," he said, "if a world without it can't function normally. If we can show that a world without Quality functions abnormally, then we have shown that Quality exists, whether it's defined or not." He thereupon proceeded to subtract Quality from a description of the world as we know it.

The first casualty from such a subtraction, he said, would be the fine arts. If you can't distinguish between good and bad in the arts they disappear. There's no point in hanging a painting on the wall when the bare wall looks just as good. There's no point to symphonies when scratches from the record or hum from the record player sound just as good.

Poetry would disappear, since it seldom makes sense and has no practical value. And interestingly, comedy would vanish too. No one would understand the jokes, since the difference between humor and no humor is pure Quality.

Next he made sports disappear. Football, baseball, games of every sort would vanish. The scores would no longer be a measurement of anything meaningful, but simply empty statistics, like the number of stones in a pile of gravel. Who would attend them? Who would pay?

Next he subtracted Quality from the marketplace and predicted the changes that would take place. Since quality of flavor would be meaningless, supermarkets would carry only basic grains such as rice, cornmeal, soybeans and flour; possibly also some ungraded meat, milk for weaning infants and vitamin and mineral supplements to make up deficiencies. Alcoholic beverages, tea, coffee and tobacco would vanish. So would movies, dances, plays and parties. We would all use public transportation. We would all wear G.I. shoes.

A huge proportion of us would be out of work, but this would probably be temporary until we relocated in essential non-Quality work. Applied science and technology would be drastically changed, but pure science, mathematics, philosophy and particularly logic would be unchanged.

Phaedrus found this last to be extremely interesting. The purely intellectual pursuits were the least affected by the subtraction of Quality. If Quality were dropped, only rationality would remain unchanged. That was odd. Why would that be?

He didn't know, but he did know that by subtracting

Quality from a picture of the world as we know it, he'd revealed a magnitude of importance of this term he hadn't known was there. The world *can* function without it, but life would be so dull as to be hardly worth living. In fact it wouldn't be worth living. The term *worth* is a Quality term. Life would just be living without any values or purpose at all.

He looked back over the distance this line of thought had taken him and decided he'd certainly proved his point. Since the world obviously doesn't function normally when Quality is subtracted, Quality exists, whether it's defined or not.

After conjuring up this vision of a Qualityless world, he was soon attracted to its resemblance to a number of social situations he had already read about. Ancient Sparta came to mind, Communist Russia and her satellites. Communist China, the *Brave New World* of Aldous Huxley and the *1984* of George Orwell. He also remembered people from his own experience who would have endorsed this Qualityless world. The same ones who tried to make him quit smoking. They wanted rational reasons for his smoking and, when he didn't have any, acted very superior, as though he'd lost face or something. They had to have reasons and plans and solutions for everything. They were his own kind. The kind he was now attacking. And he searched for a long time for a suitable name to sum up just what characterized them, so as to get a handle on this Qualityless world.

It was intellectual primarily, but it wasn't just intelligence that was fundamental. It was a certain basic attitude about the way the world was, a presumptive vision that it ran according to laws—reason—and that man's improvement lay chiefly through the discovery of these laws of reason and application of them toward satisfaction of his own desires. It was this faith that held everything together. He squinted at this vision of a Qualityless world for a while, conjured up more details, thought about it, and then squinted some more and thought some more and then finally circled back to where he was before.

Squareness.

That's the look. That sums it. Squareness. When you subtract quality you get squareness. Absence of Quality is the essence of squareness.

Some artist friends with whom he had once traveled across the United States came to mind. They were Negroes, who had always been complaining about just this Qualitylessness he was describing. Square. That was their word for it. Way back long ago before the mass media had picked it up and given it national white usage they had called all that intellectual stuff square and had wanted nothing to do with it. And there had been a fantastic mismeshing of conversations and attitudes between him and them because he was such a prime example of the squareness they were talking about. The more he had tried to pin them down on what they were talking about the vaguer they had gotten. Now with this Quality he seemed to say the same thing and talk as vaguely as they did, even though what he talked about was as hard and clear and solid as any rationally defined entity he'd ever dealt with.

Quality. That's what they'd been talking about all the time. "May, will you please, kindly *dig* it," he remembered one of them saying, "and hold up on all those wonderful seven-dollar questions? If you got to ask what *is* it all the time, you'll never get time to *know.*" Soul. Quality. The same?

The wave of crystallization rolled ahead. He was seeing two worlds, simultaneously. On the intellectual side, the square side, he saw now that Quality was a cleavage term. What every intellectual analyst looks for. You take your analytic knife, put the point directly on the term Quality and just tap, not hard, gently, and the whole world splits, cleaves, right in two—hip and square, classic and romantic, technological and humanistic—and the split is clean. There's no mess. No slop. No little items that could be one way or the other. Not just a skilled break but a very lucky break. Sometimes the best analysts, working with the most obvious lines of cleavage, can tap and get nothing but a pile of trash. And yet here was Quality; a tiny, almost unnoticeable fault line; a line of illogic in our concept of the universe; and you tapped it, and the whole universe came apart, so neatly it was almost unbelievable. He wished Kant were alive. Kant would have appreciated it. That master diamond cutter. He would see. Hold Quality undefined. That was the secret.

Phaedrus wrote, with some beginning awareness that he was involved in a strange kind of intellectual suicide, "Squareness may be succinctly and yet thoroughly defined as an inability to see quality before it's been intellectually defined, that is, before it gets all chopped up into words. . . . We have proved that quality, though undefined, exists. Its existence can be seen empirically in the classroom, and can be demonstrated logically by showing that a world without it cannot exist as we know it. What remains to be seen, the thing to be analyzed, is not quality, but those peculiar habits of thought called 'squareness' that sometimes prevent us from seeing it."

Thus did he seek to turn the attack. The subject for analysis, the patient on the table, was no longer Quality, but analysis itself. Quality was healthy and in good shape. Analysis, however, seemed to have something wrong with it that prevented it from seeing the obvious.

I look back and see Chris is way behind. "Come on!" I shout.

He doesn't answer.

"Come *on!*" I shout again.

Then I see him fall sideways and sit in the grass on the side of the mountain. I leave my pack and go back down to him. The slope is so steep I have to dig my feet in sideways. When I get there he's crying.

"I hurt my ankle," he says, and doesn't look at me.

When an ego-climber has an image of himself to protect he naturally lies to protect this image. But it's disgusting to see and I'm ashamed of myself for letting this happen. Now my own willingness to continue becomes eroded by his tears and his inner sense of defeat passes to me. I sit down, live with this for a while, then, without turning away from it, pick up his backpack and say to him, "I'll carry the packs in relays. I'll take this one up to where mine is and then you stop and wait with it so we don't lose it. Then I'll take mine up farther and then come back for yours. That way you can get plenty of rest. It'll be slower, but we'll get there."

But I've done this too soon. There's still disgust and resentment in my voice which he hears and is shamed by. He shows anger, but says nothing, for fear he'll have to carry the pack again, just frowns and ignores me while I relay the packs upward. I work off the resentment at having to do this by realizing that it isn't any more work for me, actually, than the other way. It's more work in terms of reaching the top of the mountain, but that's only the nominal goal. In terms of the real goal, putting in good minutes, one after the other, it comes out the same; in fact, better. We climb slowly upward and the resentment leaves.

For the next hour we move slowly upward, I carrying the packs in relays, to where I locate the beginning trickle of a stream. I send Chris down for water in one of the pans, which he gets. When he comes back he says, "Why are we stopping here? Let's keep going."

"This is probably the last stream we'll see for a long time, Chris, and I'm tired."

"Why are you so tired?"

Is he trying to infuriate me? He's succeeding.

"I'm tired, Chris, because I'm carrying the packs. If you're in a hurry take your pack and go on up ahead. I'll catch up with you."

He looks at me with another flicker of fear, then sits down. "I don't like this," he says, almost in tears. "I *hate* this! I'm sorry I came. Why did we come here?" He's crying again, hard.

I reply, "You make me very sorry too. You better have something for lunch."

"I don't want anything. My stomach hurts."

"Suit yourself."

He goes off a distance and picks a stem of grass and puts it in his mouth. Then he buries his face in his hands. I make lunch for myself and have a short rest.

When I wake up again he's still crying. There's

nowhere for either of us to go. Nothing to do but face up to the existing situation. But I really don't know what the existing situation is.

"Chris," I say finally.

He doesn't answer.

"Chris," I repeat.

Still no answer. He finally says, belligerently, "What?"

"I was going to say, Chris, that you don't have to prove anything to me. Do you understand that?"

A real flash of terror hits his face. He jerks his head away violently.

I say, "You don't understand what I mean by that, do you?"

He continues to look away and doesn't answer. The wind moans through the pines.

I just don't know. I just don't know what it is. It isn't just YMCA egotism that's making him *this* upset. Some minor thing reflects badly on him and it's the end of the world. When he tries to do something and doesn't get it just right he blows up or goes into tears.

I settle back in the grass and rest again. Maybe it's not having answers that's defeating both of us. I don't want to go ahead because it doesn't look like any answers ahead. None behind either. Just lateral drift. That's what it is between me and him. Lateral drift, waiting for something.

Later I hear him prowl at the knapsack. I roll over and see him glaring at me. "Where's the cheese?" he says. The tone's still belligerent.

But I'm not going to give in to it. "Help yourself," I say. "I'm not waiting on you."

He digs around and finds some cheese and crackers. I give him my hunting knife to spread the cheese with. "I think what I'm going to do, Chris, is put all the heavy stuff in my pack and the light stuff in yours. That way I won't have to go back and forth with both packs."

He agrees to this and his mood improves. It seems to have solved something for him.

My pack must be about forty or forty-five pounds now, and after we've climbed for a while an equilibrium establishes itself at about one breath for each step.

We come to a rough grade and it changes to two breaths per step. At one bank it goes to four breaths per step. Huge steps, almost vertical, hanging on to roots and branches. I feel stupid because I should have planned my way around this. The aspen staves come in handy now, and Chris takes some interest in the use of his. The packs made you top-heavy and the sticks are good insurance against toppling over. You plant one foot, plant the staff, then SWING on it, up, and take three breaths, then plant the next foot, plant the staff and SWING up. . . .

I don't know if I've got any more Chautauqua left in me today. My head gets fuzzy about this time in the afternoon . . . maybe I can establish just one overview and let it go for today. . . .

Way back long ago when we first set out on this strange voyage I talked about how John and Sylvia seemed to be running from some mysterious death force that seemed to them to be embodied in technol-

ogy, and that there were many others like them. I
talked for a while about how some of the people involved in technology seemed to be avoiding it too. An
underlying reason for this trouble was that they saw it
from a kind of "groovy dimension" that was concerned
with the immediate surface of things whereas I was
concerned with the underlying form. I called John's
style romantic, mine classic. His was, in the argot of
the sixties, "hip," mine was "square." Then we started
going into this square world to see what made it tick.
Data, classifications, hierarchies, cause-and-effect and
analysis were discussed, and somewhere along there
was some talk about a handful of sand, the world of
which we're conscious, taken from the endless landscape of awareness around us. I said a process of discrimination goes to work on this handful of sand and
divides it into parts. Classical, square understanding is
concerned with the piles of sand and the nature of the
grains and the basis for sorting and interrelating them.

Phaedrus' refusal to define Quality, in terms of this
analogy, was an attempt to break the grip of the classical sand-sifting mode of understanding and find a
point of common understanding between the classic
and romantic worlds. Quality, the cleavage term between hip and square, seemed to be it. Both worlds
used the term. Both knew what it was. It was just that
the romantic left it alone and appreciated it for what it
was and the classic tried to turn it into a set of intellectual building blocks for other purposes. Now, with
the definition blocked, the classic mind was forced to
view Quality as the romantic did, undistorted by
thought structures.

I'm making a big thing out of all this, these

classical-romantic differences, but Phaedrus didn't. He wasn't really interested in any kind of fusion differences between these two worlds. He was after something else—his ghost. In the pursuit of this ghost he went on to wider meanings of Quality which drew him further and further to his end. I differ from him in that I've no intention of going on to that end. He just passed through this territory and opened it up. I intend to stay and cultivate it and see if I can get something to grow.

I think that the referent of a term that can split a world into hip and square, classic and romantic, technological and humanistic, is an entity that can unite a world already split along these lines into one. A real understanding of Quality doesn't just serve the System, or even beat it or even escape it. A real understanding of Quality *captures* the System, tames it, and puts it to work for one's own personal use, while leaving one completely free to fulfill his inner destiny.

Now that we're up high on one side of the canyon we can see back and down and across to the other side. It's as steep there as it is here—a dark mat of greenish-black pines going up to a high ridge. We can measure our progress by sighting against it at what seems like a horizontal angle.

That's all the Quality talk for today, I guess, thank goodness. I don't mind the Quality, it's just that all the classical talk about it *isn't* Quality. Quality is just the focal point around which a lot of intellectual furniture is getting rearranged.

* * *

We stop for a break and look down below. Chris's spirits seem to be better now, but I'm afraid it's the ego thing again.

"Look how far we've come," he says.

"We've got a lot farther to go."

Later on Chris shouts to hear his echo, and throws rocks down to see where they fall. He's starting to get almost cocky, so I step up the equilibrium to where I breathe at a good swift rate, about one-and-a-half times our former speed. This sobers him somewhat and we keep on climbing.

By about three in the afternoon my legs start to get rubbery and it's time to stop. I'm not in very good shape. If you go on after that rubbery feeling you start to pull muscles and the next day is agony.

We come to a flat spot, a large knoll protruding from the side of the mountain. I tell Chris this is it for today. He seems satisfied and cheerful; maybe some progress has been made with him after all.

I'm ready for a nap, but clouds have formed in the canyon that appear ready to drop rain. They've filled in the canyon so that we can't see the bottom and can just barely see the ridge on the other side.

I break open the packs and get the tent halves out, Army ponchos, and snap them together. I take a rope and tie it between the two trees, then throw the shelter halves over it. I cut some stakes out of shrubs with the machete, and pound them in, then dig a small trench with the flat end of the machete around the tent to drain away any rainwater. We've just got everything inside when the first rain comes down.

Chris is in high spirits about the rain. We lie on our backs on the sleeping bags and watch the rain come down and hear its popping sound on the tent. The forest has a misty appearance and we both become contemplative and watch the leaves of the shrubs jolt when struck by raindrops and jolt a little ourselves when a clash of thunder comes down but feel happy that we're dry when everything around us is wet.

After a while I reach into my pack for the paperback by Thoreau, find it and have to strain a little to read it to Chris in the grey rainy light. I guess I've explained that we've done this with other books in the past, advanced books that he wouldn't normally understand. What happens is I read a sentence, he comes up with a long series of questions about it and then, when he's satisfied, I read the next sentence.

We do this with Thoreau for a while, but after half an hour I see to my surprise and disappointment that Thoreau isn't coming through. Chris is restless and so am I. The language structure is wrong for the mountain forest we're in. At least that's my feeling. The book seems tame and cloistered, something I'd never have thought of Thoreau, but there it is. He's talking to another situation, another time, just discovering the evils of technology rather than discovering the solution. He isn't talking to us. Reluctantly I put the book away again and we're both silent and meditative. It's just Chris and me and the forest and the rain. No books can guide us anymore.

Pans we've set outside the tent begin to fill up with rainwater, and later, when we have enough, we pour it all together in a pot and add some cubes of chicken bouillon and heat it over a small Sterno stove. Like any food or drink after a hard climb in the mountains, it tastes good.

Chris says, "I like camping with you better than with the Sutherlands."

"The circumstances are different," I say.

When the bouillon is gone I get out a can of pork and beans and empty it into the pot. It takes a long time to heat up, but we're in no hurry.

"It smells good," Chris says.

The rain has stopped and just occasional drops pat down on the tent.

"I think tomorrow'll be sunny," I say.

We pass the pot of pork and beans back and forth, eating from opposite sides.

"Dad, what do you think about all the time? You're always thinking all the time."

"Ohhhhhh . . . all kinds of things."

"What about?"

"Oh, about the rain, and about troubles that can happen and about things in general."

"What things?"

"Oh, about what it's going to be like for you when you grow up."

He's interested. "What's it going to be like?"

But there's a slight ego gleam in his eyes as he asks this and the answer as a result comes out masked. "I don't know," I say, "it's just what I think about."

"Do you think we'll get to the top of this canyon by tomorrow?"

"Oh yes, we're not far from the top."

"In the morning?"

"I think so."

Later he is asleep, and a damp night wind comes down from the ridge causing a sighing sound from the pines. The silhouettes of the treetops move gently with the wind. They yield and then return, then with a sigh yield and return again, restless from forces that are not part of their nature. The wind causes a flutter of one side of the tent. I get up and peg it down, then walk on the damp spongy grass of the knoll for a while, then crawl back into the tent and wait for sleep.

19

A mat of sunlit pine needles by my face slowly tells me where I am and helps dispel a dream.

In the dream I was standing in a white-painted room looking at a glass door. On the other side was Chris and his brother and mother. Chris was waving at me from the other side of the door and his brother was smiling, but his mother had tears in her eyes. Then I saw that Chris's smile was fixed and artificial and actually there was deep fear.

I moved toward the door and his smile became better. He motioned for me to open it. I was about to open it, but then didn't. His fear came back and I turned and walked away.

It's a dream that has occurred often before. Its meaning is obvious and fits some thoughts of last night. He's trying to relate to me and is afraid he never will. Things are getting clearer up here.

Beyond the flap of the tent now the needles on the

ground send vapors of mist up toward the sun. The air feels moist and cool, and while Chris still sleeps I get out of the tent carefully, stand up and stretch.

My legs and back are stiff but not painful. I do calisthenics for a few minutes to loosen them up, then sprint from the knoll into the pines. That feels better.

The pine odor is heavy and moist this morning. I squat and look down at the morning mists in the canyon below.

Later I return to the tent where a noise indicates that Chris is awake, and when I look inside I see his face stare around silently. He's a slow waker and it'll be five minutes before his mind warms up to the point where he can speak. Now he squints into the light.

"Good morning," I say.

No answer. A few raindrops fall down from the pines.

"Did you sleep well?"

"No."

"That's too bad."

"How come you're up so early?" he asks.

"It's not early."

"What time is it?"

"Nine o'clock," I say.

"I bet we didn't go to sleep until three."

Three? If he stayed awake he's going to pay for it today.

"Well, *I* got to sleep," I say.

He looks at me strangely. "You kept *me* awake."

"Me?"

Talking."

"In my sleep, you mean."

"No, about the *mountain!*"

Something is odd here. "I don't know anything about a mountain, Chris."

"Well, you talked all night about it. You said at the top of the mountain we'd see everything. You said you were going to meet me there."

I think he's been dreaming. "How could I meet you there when I'm already with you?"

"I don't know. *You* said it." He looks upset. "You sounded like you were drunk or something."

He's still half asleep. I'd better let him wake up peacefully. But I'm thirsty and remember I left the canteen behind, thinking we'd find enough water as we traveled. Dumb. There'll be no breakfast now until we're up over the ridge and far enough down to the other side where we can find a spring. "We'd better pack up and go," I say, "if we're going to get some water for breakfast." It's already warm out and probably will be hot this afternoon.

The tent comes down easily and I'm pleased to see that everything stayed dry. In a half-hour we're packed. Now except for beaten-down grass the area looks as if no one had been here.

We still have a lot of climbing to do, but on the trail we discover it's easier than yesterday. We're getting to the rounded upper portion of the ridge and the slope isn't as steep. It looks as though the pines have never been cut here. All direct light is shut out from the forest floor and there's no underbrush at all. Just a springy floor of needles that's open and spacious and easy hiking. . . .

Time to get on with the Chautauqua and the second wave of crystallization, the metaphysical one.

This was brought about in response to Phaedrus' wild meanderings about Quality when the English faculty at Bozeman, informed of their squareness, presented him with a reasonable question: "Does this undefined 'quality' of yours exist in the things we observe?" they asked. "Or is it subjective, existing only in the observer?" It was a simple, normal enough question, and there was no hurry for an answer.

Hah. There was no need for hurry. It was a finisher-offer, a knockdown question, a haymaker, a Saturday-night special—the kind you don't recover from.

Because if Quality exists in the object, then you must explain just why scientific instruments are unable to detect it. You must suggest instruments that will detect it, or live with the explanation that instruments don't detect it because your whole Quality concept, to put it politely, is a large pile of nonsense.

On the other hand, if Quality is subjective, existing only in the observer, then this Quality that you make so much of is just a fancy name for whatever you like.

What Phaedrus had been presented with by the faculty of the English department of Montana State College was an ancient logical construct known as a dilemma. A *dilemma,* which is Greek for "two premises," has been likened to the front end of an angry and charging bull.

If he accepted the premise that Quality was objective, he was impaled on one horn of the dilemma. If he accepted the other premise that Quality was subjective, he was impaled on the other horn. Either Quality is objective or subjective, therefore he was impaled no matter how he answered.

He noticed that from a number of faculty members he was receiving some good-natured smiles.

Phaedrus, however, because of his training in logic, was aware that every dilemma affords not two but three classic refutations, and he also knew of a few that weren't so classic, so he smiled back. He could take the left horn and refute the idea that objectivity implied scientific detectability. Or, he could take the right horn, and refute the idea that subjectivity implies "anything you like." Or he could go between the horns and deny that subjectivity and objectivity are the only choices. You may be sure he tested out all three.

In addition to these three classical logical refutations there are some illogical, "rhetorical" ones. Phaedrus, being a rhetorician, had these available too.

One may throw sand in the bull's eyes. He had already done this with his statement that lack of knowledge of what Quality is constitutes incompetence. It's an old rule of logic that the competence of a speaker has no relevance to the truth of what he says, and so talk of incompetence was pure sand. The world's biggest fool can say the sun is shining, but that doesn't make it dark out. Socrates, that ancient enemy of rhetorical argument, would have sent Phaedrus flying for this one, saying, "Yes, I accept your premise that I'm incompetent on the matter of Quality. Now please show an incompetent old man what Quality is. Otherwise, how am I to improve?" Phaedrus would have been allowed to stew around for a few minutes, and then been flattened with questions that proved he didn't know what Quality was either and was, by his own standards, incompetent.

One may attempt to sing the bull to sleep. Phaedrus could have told his questioners that the answer to this dilemma was beyond his humble powers of solution, but the fact that he couldn't find an answer was no logical proof that an answer couldn't be found. Wouldn't they, with their broader experience, try to help him find this answer? But it was way too late for lullabies like that. They could simply have replied, "No, we're way too square. And until you do come up with an answer, stick to the syllabus so that we don't have to flunk out your mixed-up students when we get them next quarter."

A third rhetorical alternative to the dilemma, and the best one in my opinion, was to *refuse to enter the arena*. Phaedrus could simply have said, "The attempt to classify Quality as subjective or objective is an attempt to define it. I have already said it is *undefinable*," and left it at that. I believe DeWeese actually counseled him to do this at the time.

Why he chose to disregard this advice and chose to respond to this dilemma logically and dialectically rather than take the easy escape of mysticism, I don't know. But I can guess. I think first of all that he felt the whole Church of Reason was irreversibly *in* the arena of logic, that when one put oneself outside logical disputation, one put oneself outside any academic consideration whatsoever. Philosophical mysticism, the idea that truth is indefinable and can be apprehended only by nonrational means, has been with us since the beginning of history. It's the basis of Zen practice. But it's not an academic subject. The academy, the Church of Reason, is concerned exclusively with those things that *can* be defined, and if one wants to be a mystic, his place is in a monastery, not a University. Universities are places where things should be spelled out.

I think a second reason for his decision to enter the arena was an egoistic one. He knew himself to be a pretty sharp logician and dialectician, took pride in this and looked upon this present dilemma as a challenge to his skill. I think now that trace of egotism may have been the beginning of all his troubles.

I see a deer move about two hundred yards ahead and above us through the pines. I try to point it out to Chris, but by the time he looks it's gone.

The first horn of Phaedrus' dilemma was, If Quality exists in the object, why can't scientific instruments detect it?

This horn was the mean one. From the start he saw how deadly it was. If he was going to presume to be some superscientist who could see in objects Quality that no scientist could detect, he was just proving himself to be a nut or a fool or both. In today's world, ideas that are incompatible with scientific knowledge don't get off the ground.

He remembered Locke's statement that no object, scientific or otherwise, is knowable except in terms of its qualities. This irrefutable truth seemed to suggest that the reason scientists cannot detect Quality *in* objects is because Quality is *all* they detect. The "object" is an intellectual construct *deduced* from the qualities. This answer, if valid, certainly smashed the first horn of the dilemma, and for a while excited him greatly.

But it turned out to be false. The Quality that he and

the students had been seeing in the classroom was completely different from the qualities of color or heat or hardness observed in the laboratory. Those physical properties were all measurable with instruments. His Quality—"excellence," "worth," "goodness"—was not a physical property and was not measurable. He had been thrown off by an ambiguity in the term *quality*. He wondered why that ambiguity should exist, made a mental note to do some digging into the historic roots of the word *quality,* then put it aside. The horn of the dilemma was still there.

He turned his attention to the other horn of the dilemma, which showed more promise of refutation. He thought, So Quality is whatever you like it? It angered him. The great artists of history—Raphael, Beethoven, Michelangelo—they were all just putting out what people liked. They had no goal other than to titillate the senses in a big way. Was that it? It was angering, and what was most angering about it was that he couldn't see any immediate way to cut it up logically. So he studied the statement carefully, in the same reflective way he always studied things before attacking them.

Then he saw it. He brought out the knife and excised the one word that created the entire angering effect of that sentence. The word was "just." Why should Quality be *just* what you like? Why should "what you like" be "just"? What did "just" *mean* in this case? When separated out like this for independent examination it became apparent that "just" in this case really didn't mean a damn thing. It was a purely pejorative term, whose logical contribution to the sentence was nil. Now, with that word removed, the sentence became

"Quality is what you like," and its meaning was entirely changed. It had become an innocuous truism.

He wondered why that statement had angered him so much in the first place. It had seemed so natural. Why had it taken so long to see that what it really said was "What you like is bad, or at least inconsequential." What was behind this smug presumption that what pleased you was bad, or at least unimportant in comparison to other things? It seemed the quintessence of the squareness he was fighting. Little children were trained not to do "just what they liked" but . . . but what? . . . Of course! What *others* liked. And which others? Parents, teachers, supervisors, policemen, judges, officials, kings, dictators. All authorities. When you are trained to despise "just what you like" then, of course, you become a much more obedient servant of others—a *good* slave. When you learn not to do "just what you like" then the System loves you.

But suppose you *do* just what you like? Does that mean you're going to go out and shoot heroin, rob banks and rape old ladies? The person who is counseling you not to do "just as you like" is making some remarkable presumptions as to what is likable. He seems unaware that people may not rob banks because they have considered the consequences and decided they don't like to. He doesn't see that banks exist in the first place because they're "just what people like," namely, providers of loans. Phaedrus began to wonder how all this condemnation of "what you like" ever seemed such a natural objection in the first place.

Soon he saw there was much more to this than he had been aware of. When people said, Don't do just what you like, they didn't just mean, Obey authority. They also meant something else.

This "something else" opened up into a huge area of classic scientific belief which stated that "what you like" is unimportant because it's all composed of irrational emotions within yourself. He studied this argument for a long time, then knifed it into two smaller groups which he called scientific materialism and classic formalism. He said the two are often found associated in the same person but logically are separate.

Scientific materialism, which is commoner among lay followers of science than among scientists themselves, holds that what is composed of matter or energy and is measurable by the instruments of science is real. Anything else is unreal, or at least of no importance. "What you like" is unmeasurable, and therefore unreal. "What you like" can be a fact or it can be a hallucination. Liking does not distinguish between the two. The whole purpose of scientific method is to make valid distinctions between the false and the true in nature, to eliminate the subjective, unreal, imaginary elements from one's work so as to obtain an objective, true picture of reality. When he said Quality was subjective, to them he was just saying Quality is imaginary and could therefore be disregarded in any serious consideration of reality.

On the other hand is classic formalism, which insists that what isn't understood intellectually isn't understood at all. Quality in this case is unimportant because it's an emotional understanding unaccompanied by the intellectual elements of reason.

Of these two main sources of that epithet "just," Phaedrus felt that the first, scientific materialism, was by far the easiest to cut to ribbons. This, he knew from his earlier education, was naïve science. He went after it first, using the *reductio ad absurdum*. This form of argument rests on the truth that if the inevitable conclusions from a set of premises are absurd then it follows logically that at least one of the premises that produced them is absurd. Let's examine, he said, what follows from the premise that anything not composed of mass-energy is unreal or unimportant.

He used the number zero as a starter. Zero, originally a Hindu number, was introduced to the West by the Arabs during the Middle Ages and was unknown to the ancient Greeks and Romans. How was *that*? he wondered. Had nature so subtly hidden zero that all the Greeks and all the Romans—millions of them—couldn't find it? One would normally think that zero is right out there in the open for everyone to see. He showed the absurdity of trying to derive zero from any form of mass-energy, and then asked, rhetorically, if that meant the number zero was "unscientific." If so, did that mean that digital computers, which function exclusively in terms of ones and zeros, should be limited to just ones for scientific work? No trouble finding the absurdity here.

He then went on with other scientific concepts, one by one, showing how they could not possibly exist independently of subjective considerations. He ended up with the law of gravity, in the example I gave John and Sylvia and Chris on the first night of our trip. If subjectivity is eliminated as unimportant, he said, then the entire body of science must be eliminated with it.

This refutation of scientific materialism, however, seemed to put him in the camp of philosophic idealism—

Berkeley, Hume, Kant, Fichte, Schelling, Hegel, Bradley, Bosanquet—good company all, logical to the last comma, but so difficult to justify in "common sense" language they seemed a burden to him in his defense of Quality rather than an aid. The argument that the world was all mind might be a sound logical position but it was certainly not a sound rhetorical one. It was way too tedious and difficult for a course in freshman composition. Too "far-fetched."

At this point the whole subjective horn of the dilemma looked almost as uninspiring as the objective one. And the arguments of classical formalism, when he started to examine them, made it even worse. These were the extremely forceful arguments that you shouldn't respond to your immediate emotional impulses without considering the big rational picture.

Kids are told, "Don't spend your whole allowance for bubble gum [immediate emotional impulse] because you're going to want to spend it for something else later [big picture]." Adults are told, "This paper mill may smell awful even with the best controls [immediate emotions], but without it the economy of the whole town will collapse [big picture]." In terms of your old dichotomy, what's being said is, "Don't base your decisions on romantic surface appeal without considering classical underlying form." This was something he kind of agreed with.

What the classical formalists meant by the objection "Quality is just what you like" was that this subjective, undefined "quality" he was teaching was just romantic surface appeal. Classroom popularity contests could determine whether a composition had immediate appeal, all right, but was this *Quality*? Was Quality something that you "just see" or might it be something more subtle than that, so that you wouldn't see it at all immediately, but only after a long period of time?

The more he examined this argument the more formidable it appeared. This looked like the one that might do in his whole thesis.

What made it so ominous was that it seemed to answer a question that had arisen often in class and which he always had to answer somewhat casuistically. This was the question, If everyone knows what quality is, why is there such a disagreement about it?

His casuist answer had been that although pure *Quality* was the same for everyone, the *objects* that people said Quality *inhered* in varied from person to person. As long as he left Quality undefined there was no way to argue with this but he knew and he knew the students knew that it had the smell of falseness about it. It didn't really answer the question.

Now there was an alternative explanation: people disagreed about Quality because some just used their immediate emotions whereas others applied their overall knowledge. He knew that in any popularity contest among English teachers, this latter argument which bolstered their authority would win overwhelming endorsement.

But this argument was completely devastating. Instead of one single, uniform Quality now there appeared to be *two* qualities: a romantic one, just seeing, which the students had; and a classic one, overall understanding, which the teachers had. A hip one and a square one. Squareness was not the absence of Quality; it was classic Quality. Hipness was not just presence of Quality; it was mere romantic Quality. The hip-square cleavage he'd discovered was still there, but Quality didn't now seem to fall entirely on one side of the cleavage, as he'd previously supposed. Instead, Quality itself cleaved into two kinds, one on each side of the cleavage line. His simple, neat, beautiful, undefined Quality was starting to get complex.

He didn't like the way this was going. The cleavage term that was going to unify the classic and romantic ways of looking at things had itself been cleaved into two parts and could no longer unify anything. It had been caught in an analytic meat grinder. The knife of subjectivity-and-objectivity had cut Quality in two and killed it as a working concept. If he was going to save it, he couldn't let that knife get it.

And really, the Quality he was talking about *wasn't* classic Quality *or* romantic Quality. It was beyond both of them. And by God, it wasn't subjective or objective either, it was beyond both of *those* categories. Actually this whole dilemma of subjectivity-objectivity, of mind-matter, with relationship to Quality was unfair. That mind-matter relationship has been an intellectual hang-up for centuries. They were just putting that hang-up on top of Quality to drag Quality down. How could *he* say whether Quality was mind or matter when there was no logical clarity as to what was mind and what was matter in the first place?

And so: he rejected the left horn. Quality is not objective, he said. It doesn't reside in the material world.

Then: he rejected the right horn. Quality is not subjective, he said. It doesn't reside merely in the mind.

And finally: Phaedrus, following a path that to his knowledge had never been taken before in the history of Western thought, went straight between the horns of the subjectivity-objectivity dilemma and said Quality is neither a part of mind, nor is it a part of matter. It is a *third* entity which is independent of the two.

He was heard along the corridors and up and down the stairs of Montana Hall singing softly to himself, almost under his breath, "Holy, holy, holy . . . blessed Trinity."

And there is a faint, faint fragment of memory, possibly wrong, possibly just something I'm imagining, that says he just let the whole thought structure sit like that for weeks, without carrying it any further.

Chris shouts, "When are we going to get to the top?"

"Probably quite a way yet," I reply.

"Will we see a lot?"

"I think so. Look for blue sky between the trees. As long as we can't see sky we know it's a way yet. The light will come through the trees when we round the top."

Last night's rain has soaked this soft duff of needles sufficiently to make them good walking. Sometimes when it's really dry on a slope like this they become slippery and you have to dig your feet into them edgewise or you'll slide down.

I say to Chris, "Isn't it great when there's no underbrush like this?"

"Why isn't there any?" he asks.

"I think this area must never have been logged. When a forest is left alone like this for centuries, the trees shut out all the underbrush."

"It's like a park," Chris says. "You can sure see all around." His mood seems much better than yesterday. I think he'll be a good traveler from here on. This forest silence improves anyone.

*   *   *

The world now, according to Phaedrus, was composed of three things: mind, matter, and Quality. The fact that he had established no relationship between them didn't bother him at first. If the relationship between mind and matter had been fought over for centuries and wasn't yet resolved, why should he, in a matter of a few weeks, come up with something conclusive about Quality? So he let it go. He put it up on a kind of mental shelf where he put all kinds of questions he had no immediate answers for. He knew the metaphysical trinity of subject, object and Quality would sooner or later have to be interrelated but he was in no hurry about it. It was just so satisfying to be beyond the danger of those horns that he relaxed and enjoyed it as long as he could.

Eventually, however, he examined it more closely. Although there's no logical objection to a metaphysical trinity, a three-headed reality, such trinities are not common or popular. The metaphysician normally seeks either a monism, such as God, which explains the nature of the world as a manifestation of one single thing, or he seeks a dualism, such as mind-matter, which explains it as two things, or he leaves it as a pluralism, which explains it as a manifestation of an indefinite number of things. But three is an awkward number. Right away you want to know, Why three? What's the relationship among them? And as the need for relaxation diminished Phaedrus became curious about this relationship too.

He noted that although normally you associate Quality with objects, feelings of Quality sometimes occur without any object at all. This is what led him at first to think that maybe Quality is all subjective. But subjective pleasure wasn't what he meant by Quality either. Quality *decreases* subjectivity. Quality takes you out of yourself, makes you aware of the world around you. Quality is *opposed* to subjectivity.

I don't know how much thought passed before he arrived at this, but eventually he saw that Quality couldn't be independently related with either the subject or the object but could be found *only in the relationship of the two with each other.* It is the point at which subject and object meet.

That sounded warm.

Quality is not a *thing.* It is an *event.*

Warmer.

It is the event at which the subject becomes aware of the object.

And because without objects there can be no subject—because the objects create the subject's awareness of himself—Quality is the event at which awareness of both subjects and objects is made possible.

Hot.

Now he knew it was coming.

This means Quality is not just the *result* of a collision between subject and object. The very existence of subject and object themselves is *deduced* from the Quality event. The Quality event is the *cause* of the subjects and objects, which are then mistakenly presumed to be the cause of the Quality!

Now he had that whole damned evil dilemma by the throat. The dilemma all the time had this unseen vile presumption in it, for which there was no logical justification, that Quality was the *effect* of subjects and objects. It was *not*! He brought out his knife.

"The sun of quality," he wrote, "does not revolve around the subjects and objects of our existence. It does not just passively illuminate them. It is not subordinate to them in any way. It has *created* them. They are subordinate to *it*!"

And at that point, when he wrote that, he knew he had reached some kind of culmination of thought he had been unconsciously striving for over a long period of time.

"Blue sky!" shouts Chris.

There it is, way above us, a narrow patch of blue through the trunks of the trees.

We move faster and the patches of blue become larger and larger through the trees and soon we see that the trees thin out to a bare spot at the summit. When the summit is about fifty yards away I say, "Let's go!" and start to dash for it, throwing into the effort all the reserves of energy I've been saving.

I give it everything I have, but Chris gains on me. Then he passes me, giggling. With the heavy load and high altitude we're not setting any records but now we're just charging up with all we have.

Chris gets there first, while I just break out of the trees. He raises his arms and shouts, "The Winner!"

Egotist.

I'm breathing so hard when I arrive I can't speak. We just drop our packs from our shoulders and lie down against some rocks. The crust of the ground is dry from the sun, but underneath is mud from last night's rain. Below us and miles away beyond the forested slopes and the fields beyond them is the Gallatin Valley. At one corner of the valley is Bozeman. A grasshopper jumps up from the rock and soars down and away from us over the trees.

"We made it," Chris says. He is very happy. I am still too winded to answer. I take off my boots and socks which are soggy with sweat and set them out to dry on a rock. I stare at them meditatively as vapors from them rise up toward the sun.

20

Evidently I've slept. The sun is hot. My watch says a few minutes before noon. I look over the rock I'm leaning against and see Chris sound asleep on the other side. Way up above him the forest stops and barren grey rock leads into patches of snow. We can climb the back of this ridge straight up there, but it would be dangerous toward the top. I look up at the top of the mountain for a while. What was it Chris said I told him last night?—"I'll see you at the top of mountain" . . . no . . . "I'll *meet* you at the top of the mountain."

How could I meet him at the top of the mountain when I'm already with him? Something's very strange about that. He said I told him something else too, the other night—that it's lonely here. That contradicts what I actually believe. I don't think it's lonely here at all.

A sound of falling rock draws my attention over to one side of the mountain. Nothing moves. Completely still.

It's all right. You hear little rockslides like this all the time.

Not so little sometimes, though. Avalanches start with little slides like that. If you're above them or beside them, they're interesting to watch. But if they're above you—no help then. You just have to watch it come.

People say strange things in their sleep, but why would I tell him I'll *meet* him? And why would he think I was awake? There's something really wrong there that produces a very bad quality feeling, but I don't know what it is. First you get the feeling, then you figure out why.

I hear Chris move and turn and see him look around.

"Where are we?" he asks.

"Top of the ridge."

"Oh," he says. He smiles.

I break open a lunch of Swiss cheese, pepperoni and crackers. I cut up the cheese and then the pepperoni in careful, neat slices. The silence allows you to do each thing right.

"Let's build a cabin here," he says.

"Ohhhhh," I groan, "and climb up to it every day?"

"Sure," he teases. "That wasn't hard."

Yesterday is long ago in his memory. I pass some cheese and crackers over to him.

"What are you always thinking about?" he asks.

"Thousands of things," I answer.

"What?"

"Most of them wouldn't make any sense to you."

"Like what?"

"Like why I told you I'd meet you at the top of the mountain."

"Oh," he says, and looks down.

"You said I sounded drunk," I tell him.

"No, not drunk," he says, still looking down. The way he looks away from me makes me wonder all over again if he's telling the truth.

"How then?"

He doesn't answer.

"How then, Chris?"

"Just different."

"How?"

"Well, *I* don't *know*!" He looks up at me and there's a flicker of fear. "Like you used to sound a long time ago," he says, and then looks down.

"When?"

"When we lived here."

I keep my face composed so that he sees no change of expression in it, then carefully get up and go over and methodically turn the socks on the rock. They've dried long ago. As I return with them I see his glance is still on me. Casually I say, "I didn't know I sounded different."

He doesn't reply to this.

I put the socks on and slip the boots over them.

"I'm thirsty," Chris says.

"We shouldn't have too far to go down to find water," I say, standing up. I look at the snow for a while, then say, "You ready to go?"

He nods and we get the packs on.

As we walk along the summit toward the beginning of a ravine we hear another clattering sound of falling rock, much louder than the first one I heard just a while ago. I look up to see where it is. Still nothing.

"What was that?" Chris asks.

"Rockslide."

We both stand still for a moment, listening. Chris asks, "Is there somebody up there?"

"No, I think it's just melting snow that's loosening stones. When it's really hot like this in the early part of the summer you hear a lot of small rockslides. Sometimes big ones. It's part of the wearing down of the mountains."

"I didn't know mountains wore out."

"Not wore out, wore *down*. They get rounded and gentle. These mountains are still unworn."

Everywhere around us now, except above, the sides of the mountain are covered with the blackish green of the forest. In the distance the forest looks like velvet.

I say, "You look at these mountains now, and they look so permanent and peaceful, but they're changing all the time and the changes aren't always peaceful. Underneath us, beneath us here right now, there are forces that can tear this whole mountain apart."

"Do they ever?"

"Ever what?"

"Tear the whole mountain apart?"

"Yes," I say. Then I remember: "Not far from here there are nineteen people lying dead under millions of tons of rock. Everyone was amazed there were only nineteen."

"What happened?"

"They were just tourists from the east who had stopped for the night at a campground. During the night the underground forces broke free and when the rescuers saw what had happened the next morning, they just shook their heads. They didn't even try to excavate. All they could have done was dig down through hundreds of feet of rock for bodies that would just have to be buried all over again. So they left them there. They're still there now."

"How did they know there were nineteen?"

"Neighbors and relatives from their hometowns reported them missing."

Chris stares at the top of the mountain before us. "Didn't they get any warning?"

"I don't know."

"You'd think there'd be a warning."

"Maybe there was."

We walk to where the ridge we are on creases inward to the start of a ravine. I see that we can follow this ravine down and eventually find water in it. I start angling down now.

Some more rocks clatter up above. Suddenly I'm frightened.

"Chris," I say.

"What?"

"You know what I think?"

"No, what?"

"I think we'd be very smart if we let that mountaintop go for now and try it another summer."

He's silent. Then he says, "Why?"

"I have bad feelings about it."

He doesn't say anything for a long time. Finally he says, "Like what?"

"Oh, I just think that we could get caught up there in a storm or a slide or something and we'd be in real trouble."

More silence. I look up and see real disappointment

in his face. I think he knows I'm leaving something out. "Why don't you think about it," I say, "and then when we get to some water and have lunch we'll decide."

We continue walking down. "Okay?" I say.

He finally says, "Okay," in a noncommittal voice.

The descent is easy now but I see it will be steeper soon. It's still open and sunny here but soon we'll be in trees again.

I don't know what to make of all this weird talk at night except that it's not good. For either of us. It sounds like all the strain of this cycling and camping and Chautauqua and all these old places has a bad effect on me that appears at night. I want to clear out of here as fast as possible.

I don't suppose that sounds like the old days to Chris either. I spook very easily these days, and am not ashamed to admit it. *He* never spooked at anything. Never. That's the difference between us. That's why I'm alive and he's not. If he's up there, some psychic entity, some ghost, some Doppelgänger waiting up there for us in God knows what fashion . . . well, he's going to have to wait a long time. A very long time.

These damned heights get eerie after a while. I want to go down, way down; far, far down.

To the ocean. That sounds right. Where the waves roll in slowly and there's always a roar and you can't fall anywhere. You're already there.

Now we enter the trees again, and the sight of the mountaintop is obscured by their branches and I'm glad.

I think we've gone as far along Phaedrus' path as we want to go in this Chautauqua too. I want to leave his path now. I've given him all due credit for what he thought and said and wrote, and now I want to develop on my own some of the ideas he neglected to pursue. The title of this Chautauqua is "Zen and the Art of Motorcycle Maintenance," not "Zen and the Art of Mountain Climbing," and there are no motorcycles on the tops of mountains, and in my opinion very little Zen. Zen is the "spirit of the valley," not the mountaintop. The only Zen you find on the tops of mountains is the Zen you bring up there. Let's get out of here.

"Feels good to be going down, doesn't it?" I say.

No answer.

We're going to have a little fight, I'm afraid.

You go up to the mountaintop and all you're gonna get is a great big heavy stone tablet handed to you with a bunch of rules on it.

That's about what happened to him.

Thought he was a goddamned Messiah.

Not *me,* boy. The hours are way too long, and the pay is *way* too short. Let's go. Let's go. . . .

Soon I'm clomping down the slope in a kind of two-step idiot gallop . . . *ga-dump, ga-dump, ga-dump* . . . until I hear Chris holler, "SLOW DOWN!" and see he is a couple of hundred yards back through the trees.

So I slow down, but after a while see he is deliberately lagging behind. He's disappointed, of course.

I suppose what I ought to do in the Chautauqua is just point out in summary form the direction Phaedrus went, without evaluation, and then get on with my own thing. Believe me, when the world is seen not as a duality of mind and matter but as a trinity of quality, mind, and matter, then the art of motorcycle maintenance and other arts take on a dimension of meaning they never had. The specter of technology the Sutherlands are running from becomes not an evil but a positive fun thing. And to demonstrate that will be a long fun task.

But first, to give this other specter his walking papers, I should say the following:

Perhaps he would have gone in the direction I'm now about to go in if this second wave of crystallization, the metaphysical wave, had finally grounded out where I'll be grounding it out, that is, in the everyday world. I think metaphysics is good if it improves everyday life; otherwise forget it. But unfortunately for him it didn't ground out. It went into a third mystical wave of crystallization from which he never recovered.

He'd been speculating about the relationship of Quality to mind and matter and had identified Quality as the parent of mind and matter, that event which gives birth to mind and matter. This Copernican inversion of the relationship of Quality to the objective world could sound mysterious if not carefully explained, but he didn't mean it to be mysterious. He simply meant that at the cutting edge of time, before an object can be distinguished, there must be a kind of nonintellectual awareness, which he called awareness of Quality. You can't be aware that you've seen a tree until *after* you've seen the tree, and between the instant of vision and instant of awareness there must be a time lag. We sometimes think of that time lag as unimportant. But there's no justification for thinking that the time lag is unimportant—none *whatsoever.*

The past exists only in our memories, the future only

in our plans. The present is our only reality. The tree that you are aware of intellectually, because of that small time lag, is always in the past and therefore is always unreal. *Any* intellectually conceived object is *always* in the past and therefore *unreal.* Reality is always the moment of vision *before* the intellectualization takes place. There is no other reality. This preintellectual reality is what Phaedrus felt he had properly identified as Quality. Since all intellectually identifiable things must emerge *from* this preintellectual reality, Quality is the *parent,* the *source* of all subjects and objects.

He felt that intellectuals usually have the greatest trouble seeing this Quality, precisely because they are so swift and absolute about snapping everything into intellectual form. The ones who have the easiest time seeing this Quality are small children, uneducated people and culturally "deprived" people. These have the least predisposition toward intellectuality from cultural sources and have the least formal training to instill it further into them. That, he felt, is why squareness is such a uniquely intellectual disease. He felt he'd been accidentally immunized from it, or at least to some extent broken from the habit by his failure from school. After that he felt no compulsive identification with intellectuality and could examine anti-intellectual doctrines with sympathy.

Squares, he said, because of their prejudices toward intellectuality usually regard Quality, the preintellectual reality, as unimportant, a mere uneventful transition period between objective reality and subjective perception of it. Because they have preconceived ideas of its unimportance they don't seek to find out if it's in any way different from their intellectual conception of it.

It *is* different, he said. Once you begin to hear the sound of that Quality, see that Korean wall, that nonintellectual reality in its pure form, you want to forget all that word stuff, which you finally begin to see is always somewhere else.

Now, armed with his new time-interrelated metaphysical trinity, he had that romantic-classic Quality split, the one which had threatened to ruin him, completely stopped. They couldn't cut up Quality now. He could sit there and at his leisure cut *them* up. Romantic Quality always correlated with instantaneous impressions. Square Quality always involved multiple considerations that extended over a period of time. Romantic Quality was the present, the here and now of things. Classic Quality was always concerned with more than just the present. The relation of the present to the past and future was always considered. If you conceived the past and future to be all contained in the present, why, that was groovy, the present was what you lived for. And if your motorcycle is working, why worry about it? But if you consider the present to be merely an instant between the past and the future, just a passing moment, then to neglect the past and future for the present is bad Quality indeed. The motorcycle may be working now, but when was the oil level last checked? Fuss-budgetry from the romantic view, but good common sense from the classic.

Now we had two different kinds of Quality but they no longer split Quality itself. They were just two different time aspects of Quality, short and long. What had previously been asked for was a metaphysical hierarchy that looked like this:

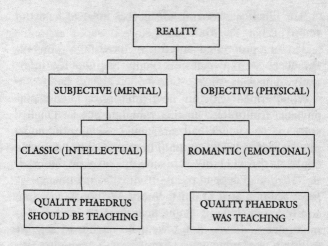

What he gave them in return was a metaphysical hierarchy that looked like this:

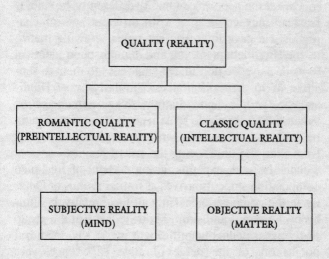

The Quality he was teaching was not just a part of reality, it was the whole thing.

He then proceeded in terms of the trinity to answer the question, Why does everybody see Quality differently? This was the question he had always had to answer speciously before. Now he said, "Quality is shapeless, formless, indescribable. To see shapes and forms is to intellectualize. Quality is independent of any such shapes and forms. The names, the shapes and forms we give Quality depend only partly on the Quality. They also depend partly on the *a priori* images we have accumulated in our memory. We constantly seek to find, in the Quality event, analogues to our previous experiences. If we didn't we'd be unable to act. We build up our language in terms of these analogues. We build up our whole culture in terms of these analogues."

The reason people see Quality differently, he said, is because they come to it with different sets of analogues. He gave linguistic examples, showing that to us the Hindi letters *da, da,* and *dha* all sound identical to us because we don't have analogues to them to sensitize us to their differences. Similarly, most Hindi-speaking people cannot distinguish between *da* and *the* because they are not so sensitized. It is not uncommon, he said, for Indian villagers to see ghosts. But they have a terrible time seeing the law of gravity.

This, he said, explains why a classful of freshman composition students arrives at similar ratings of Quality in the compositions. They all have relatively similar backgrounds and similar knowledge. But if a group of foreign students were brought in, or, say, medieval poems out of the range of class experience were brought in, then the students' ability to rank Quality would probably not correlate as well.

In a sense, he said, it's the student's choice of Quality that defines *him*. People differ about Quality, not because Quality is different, but because people are different in terms of experience. He speculated that if two people had identical *a priori* analogues they would see Quality identically every time. There was no way to test this, however, so it had to remain just speculation.

In answer to his colleagues at school he wrote:

"Any philosophic explanation of Quality is going to be both false and true precisely because it is a philosophic explanation. The process of philosophic explanation is an analytic process, a process of breaking something down into subjects and predicates. What I mean (and everybody else means) by the word *quality* cannot be broken down into subjects and predicates. This is not because Quality is so mysterious but because Quality is so simple, immediate and direct.

"The easiest intellectual analogue of pure Quality that people in our environment can understand is that 'Quality is the response of an organism to its environment' [he used this example because his chief questioners seemed to see things in terms of stimulus-response behavior theory]. An amoeba, placed on a plate of water with a drip of dilute sulfuric acid placed nearby, will pull away from the acid (I think). If it could speak the amoeba, without knowing anything about sulfuric acid, could say, 'This environment has poor quality.' If it had a nervous system it would act in a much more complex way to overcome the poor quality of the environment. It would seek analogues, that is, images and symbols from its previous experience, to define the unpleasant nature of its new environment and thus 'understand' it.

"In our highly complex organic state we advanced organisms respond to our environment with an invention of many marvelous analogues. We invent earth and heavens, trees, stones and oceans, gods, music, arts, language, philosophy, engineering, civilization and science. We call these analogues reality. And they *are* reality. We mesmerize our children in the name of truth into knowing that they *are* reality. We throw anyone who does not accept these analogues into an insane asylum. But that which causes us to invent the analogues is Quality. Quality is the continuing stimulus which our environment puts upon us to create the world in which we live. All of it. Every last bit of it.

"Now, to take that which has caused us to create the world, and include it within the world we have created, is clearly impossible. That is why Quality cannot be defined. If we do define it we are defining something less than Quality itself."

I remember this fragment more vividly than any of the others, possibly because it is the most important of all. When he wrote it he felt momentary fright and was about to strike out the words "All of it. Every last bit of it." Madness there. I think he saw it. But he couldn't see any logical reason to strike these words out and it was too late now for faintheartedness. He ignored his warning and let the words stand.

He put his pencil down and then . . . felt something let go. As though something internal had been strained too hard and had given way. Then it was too late.

He began to see that he had shifted away from his

original stand. He was no longer talking about a meta-physical trinity but an absolute monism. Quality was the source and substance of everything.

A whole new flood of philosophic associations came to mind. Hegel had talked like this, with his Absolute Mind. Absolute Mind was independent too, both of objectivity and subjectivity.

However, Hegel said the Absolute Mind was the source of everything, but then excluded romantic experience from the "everything" it was the source of. Hegel's Absolute was completely classical, completely rational and completely orderly.

Quality was not like that.

Phaedrus remembered Hegel had been regarded as a bridge between Western and Oriental philosophy. The Vedanta of the Hindus, the Way of the Taoists, even the Buddha had been described as an absolute monism similar to Hegel's philosophy. Phaedrus doubted at the time, however, whether mystical Ones and metaphysical monisms were interconvertible since mystical Ones follow no rules and metaphysical monisms do. His Quality was a metaphysical entity, not a mystic one. Or was it? What was the difference?

He answered himself that the difference was one of definition. Metaphysical entities are defined. Mystical Ones are not. That made Quality mystical. No. It was really both. Although he'd thought of it purely in philosophical terms up to now as metaphysical, he had all along refused to define it. That made it mystic too. Its indefinability freed it from the rules of metaphysics.

Then, on impulse, Phaedrus went over to his bookshelf and picked out a small, blue, cardboard-bound book. He'd hand-copied this book and bound it himself years before, when he couldn't find a copy for sale anywhere. It was the 2,400-year-old *Tao Te Ching* of Lao Tzu. He began to read through the lines he had read many times before, but this time he studied it to see if a certain substitution would work. He began to read and interpret it at the same time.

He read:

> *The quality that can be defined is not the Absolute Quality.*

That was what he had said.

> *The names that can be given it are not Absolute names.*
> *It is the origin of heaven and earth.*
> *When named it is the mother of all things. . . .*

Exactly.

> *Quality* [romantic Quality] *and its manifestations* [classic Quality] *are in their nature the same. It is given different names* [subjects and objects] *when it becomes classically manifest.*
> *Romantic quality and classic quality together may be called the "mystic."*
> *Reaching from mystery into deeper mystery, it is the gate to the secret of all life.*
> *Quality is all-pervading.*
> *And its use is inexhaustible!*
> *Fathomless!*
> *Like the fountainhead of all things . . .*

Yet crystal clear like water it seems to remain.
I do not know whose Son it is.
An image of what existed before God.
. . . Continuously, continuously it seems to
* remain. Draw upon it and it serves you with*
* ease . . .*
Looked at but cannot be seen . . . listened to but
* cannot be heard . . . grasped at but cannot be*
* touched . . . these three elude all our*
* inquiries and hence blend and become one.*
Not by its rising is there light,
Not by its sinking is there darkness
Unceasing, continuous
It cannot be defined
And reverts again into the realm of nothingness
That is why it is called the form of the formless
The image of nothingness
That is why it is called elusive
Meet it and you do not see its face
Follow it and you do not see its back
He who holds fast to the quality of old
Is able to know the primeval beginnings
Which are the continuity of quality.

Phaedrus read on through line after line, verse after verse of this, watched them match, fit, slip into place. Exactly. *This* was what he meant. *This* was what he'd been saying all along, only poorly, mechanistically. There was nothing vague or inexact about this book. It was as precise and definite as it could be. It was what he had been saying, only in a different language with different roots and origins. He was from another valley seeing what was in *this* valley, not now as a story told by strangers but as a part of the valley he was from. He was seeing it all.

He had broken the code.

He read on. Line after line. Page after page. Not a discrepancy. What he had been talking about all the time as Quality was here the Tao, the great central generating force of all religions, Oriental and Occidental, past and present, all knowledge, everything.

Then his mind's eye looked up and caught his own image and realized where he was and what he was seeing and . . . I don't know what really happened . . . but now the slippage that Phaedrus had felt earlier, the internal parting of his mind, suddenly gathered momentum, as do the rocks at the top of a mountain. Before he could stop it, the sudden accumulated mass of awareness began to grow and grow into an avalanche of thought and awareness out of control; with each additional growth of the downward tearing mass loosening hundreds of times its volume, and then that mass uprooting hundreds of times its volume more, and then hundreds of times that; on and on, wider and broader; until there was nothing left to stand.

No more anything.

It all gave way from under him.

21

"You're not very brave, are you?" Chris says.

"No," I answer, and pull the rind of a slice of salami between my teeth to remove the meat. "But you'd be astonished at how smart I am."

We're down quite a way from the summit now, and the mixed pines and leafy underbrush are much higher here and more closed in than they were at this altitude on the other side of the canyon. Evidently more rain gets into this canyon. I gulp down a large quantity of water from a pot Chris has filled at the stream here, then look at him. I can see by his expression that he's resigned himself to going down and there's no need to lecture him or argue. We finish the lunch off with a part of a bag of candy, wash it down with another pot of water and lay back on the ground for a rest. Mountain springwater has the best taste in the world.

After a while Chris says, "I can carry a heavier load now."

"Are you sure?"

"Sure I'm sure," he says, a little haughtily.

Gratefully I transfer some of the heavier stuff to his pack and we put the packs on, wriggling through the shoulder straps on the ground and then standing up. I can feel the difference in weight. He can be considerate when he's in the mood.

From here on it looks like a slow descent. This slope has evidently been logged and there's a lot of underbrush higher than our heads that makes it slow going. We'll have to work our way around it.

What I want to do now in the Chautauqua is get away from intellectual abstractions of an extremely general nature and into some solid, practical, day-to-day information, and I'm not quite sure how to go about this.

One thing about pioneers that you don't hear mentioned is that they are invariably, by their nature, messmakers. They go forging ahead, seeing only their noble, distant goal, and never notice any of the crud and debris they leave behind them. Someone else gets to clean that up and it's not a very glamorous or interesting job. You have to depress for a while before you can get down to doing it. Then, once you have depressed into a really low-key mood, it isn't so bad.

To discover a metaphysical relationship of Quality and the Buddha at some mountaintop of personal experience is very spectacular. And very unimportant. If that were all this Chautauqua was about I should be dismissed. What's important is the relevance of such a discovery to all the valleys of this world, and all the dull, dreary jobs and monotonous years that await all of us in them.

Sylvia knew what she was talking about the first day

when she noticed all those people coming the other way. What did she call it? A "funeral procession." The task now is to get back down to that procession with a wider kind of understanding than exists there now.

First of all I should say that I don't know whether Phaedrus' claim that Quality is the Tao is true. I don't know of any way of testing it for truth, since all he did was simply compare his understanding of one mystic entity with another. He certainly thought they were the same, but he may not have completely understood what Quality was. Or, more likely, he may not have understood the Tao. He certainly was no sage. And there's plenty of advice for sages in that book he would have done well to heed.

I think, furthermore, that all his metaphysical mountain climbing did absolutely nothing to further either our understanding of what Quality is or of what the Tao is. Not a thing.

That sounds like an overwhelming rejection of what he thought and said, but it isn't. I think it's a statement he would have agreed with himself, since any description of Quality is a kind of definition and must therefore fall short of its mark. I think he might even have said that statements of the kind he had made, which fall short of their mark, are even *worse* than no statement at all, since they can be easily mistaken for truth and thus *retard* an understanding of Quality.

No, he did nothing for Quality or the Tao. What benefited was reason. He showed a way by which reason may be *expanded* to include elements that have previously been unassimilable and thus have been considered irrational. I think it's the overwhelming presence of these irrational elements crying for assimilation that

creates the present bad quality, the chaotic, disconnected spirit of the twentieth century. I want to go at
these now in as orderly a manner as possible.

We're on steep mucky soil now that's hard to keep a
footing in. We grab branches and shrubs to steady ourselves. I take a step, then figure where my next step
will be, then take this step, then look again.

Soon the brush becomes so thick I see we will have
to hack through it. I sit down while Chris gets the machete from the pack on my back. He hands it to me,
then, hacking and chopping, I head into the brush. It's
slow going. Two or three branches must be cut for
every step. It may go on like this for a long time.

The first step down from Phaedrus' statement that
"Quality is the Buddha" is a statement that such an as-
sertion, if true, provides a rational basis for a unifica-
tion of three areas of human experience which are now
disunified. These three areas are Religion, Art and Science. If it can be shown that Quality is the central term
of all three, and that this Quality is not of many kinds
but of one kind only, then it follows that the three disunified areas have a basis for interconversion.

The relationship of Quality to the area of Art has
been shown rather exhaustively through a pursuit of
Phaedrus' understanding of Quality in the Art of rhetoric. I don't think much more in the way of analysis
need be made there. Art is high-quality endeavor. That
is all that really needs to be said. Or, if something more
high-sounding is demanded: Art is the Godhead as revealed in the works of man. The relationship established by Phaedrus makes it clear that the two

enormously different sounding statements are actually identical.

In the area of Religion, the rational relationship of Quality to the Godhead needs to be more thoroughly established, and this I hope to do much later on. For the time being one can meditate on the fact that the old English roots for the Buddha and Quality, *God* and *good,* appear to be identical.

It's in the area of Science that I want to focus attention in the immediate future, for this is the area that most badly needs the relationship established. The dictum that Science and its offspring, technology, are "value free," that is, "quality free," has got to go. It's that "value freedom" that underlines the death-force effect to which attention was brought early in the Chautauqua. Tomorrow I intend to start on that.

For the remainder of the afternoon we climb down over grey weathered trunks of deadfalls and angle back and forth on the steep slope.

We reach a cliff, angle along its edge in search of a way down, and eventually a narrow draw appears which we're able to descend. It continues down through a rocky crevice in which there is a little rivulet. Shrubs and rocks and muck and roots of huge trees watered by the rivulet fill the crevice. Then we hear the roar of a much larger creek in the distance.

We cross the creek using a rope, which we leave behind, then on the road beyond find some other campers who give us a ride into town.

In Bozeman it's dark and late. Rather than wake up the DeWeeses and ask them to drive in, we check in at the main downtown hotel. Some tourists in the lobby stare at us. With my old Army clothes, walking stick, two-day beard and black beret I must look like some old-time Cuban revolutionary, in for a raid.

In the hotel room we exhaustedly dump everything on the floor. I empty into a waste basket the stones picked up by my boots from the rushing water of the stream, then set the boots by a cold window to dry slowly. We collapse into the beds without a word.

22

The next morning we check out of the hotel feeling re-freshed, say goodbye to the DeWeeses, and head north on the open road out of Bozeman. The DeWeeses wanted us to stay, but a peculiar itching to move west and get on with my thoughts has taken over. I want to talk today about a person whom Phaedrus never heard of, but whose writings I've studied quite extensively in preparation for this Chautauqua. Unlike Phaedrus, this man was an international celebrity at thirty-five, a liv-ing legend at fifty-eight, whom Bertrand Russell has described as "by general agreement, the most eminent scientific man of his generation." He was an as-tronomer, a physicist, a mathematician and philoso-pher all in one. His name was Jules Henri Poincaré.

It always seemed incredible to me, and still does, I guess, that Phaedrus should have traveled along a line of thought that had never been traveled before. Some-one, somewhere, must have thought of all this before, and Phaedrus was such a poor scholar it would have been just like him to have duplicated the common-places of some famous system of philosophy he hadn't taken the trouble to look into.

So I spent more than a year reading the very long and sometimes very tedious history of philosophy in a search for duplicate ideas. It was a fascinating way to read the history of philosophy, however, and a thing occurred of which I still don't know quite what to make. Philosophical systems that are supposed to be greatly opposed to one another *both* seem to be saying something very close to what Phaedrus thought, with minor variations. Time after time I thought I'd found whom he was duplicating, but each time, because of what appeared to be some slight differences, he took a greatly different direction. Hegel, for example, whom I referred to earlier, rejected Hindu systems of philosophy as no philosophy at all. Phaedrus seemed to assimilate them, or *be* assimilated by them. There was no feeling of contradiction.

Eventually I came to Poincaré. Here again there was little duplication but another kind of phenomeonon. Phaedrus follows a long and tortuous path into the highest abstractions, seems about to come down and then stops. Poincaré starts with the most basic scientific verities, works up to the same abstractions and then stops. Both trails stop *right at each other's end*! There is perfect continuity between them. When you live in the shadow of insanity, the appearance of another mind that thinks and talks as yours does is something close to a blessed event. Like Robinson Crusoe's discovery of footprints on the sand.

Poincaré lived from 1854 to 1912, a professor at the University of Paris. His beard and pince-nez were reminiscent of Henri Toulouse-Lautrec, who lived in Paris at the same time and was only ten years younger.

During Poincaré's lifetime, an alarmingly deep crisis in the foundations of the exact sciences had begun. For years scientific truth had been beyond the possibility of a doubt; the logic of science was infallible, and if the scientists were sometimes mistaken, this was assumed to be only from their mistaking its rules. The great questions had all been answered. The mission of science was now simply to refine these answers to greater and greater accuracy. True, there were still unexplained phenomena such as radioactivity, transmission of light through the "ether," and the peculiar relationship of magnetic to electric forces; but these, if past trends were any indication, had eventually to fall. It was hardly guessed by anyone that within a few decades there would be no more absolute space, absolute time, absolute substance or even absolute magnitude; that classical physics, the scientific rock of ages, would become "approximate"; that the soberest and most respected of astronomers would be telling mankind that if it looked long enough through a telescope powerful enough, what it would see was the back of its own head!

The basis of the foundation-shattering Theory of Relativity was as yet understood only by very few, of whom Poincaré, as the most eminent mathematician of his time, was one.

In his *Foundations of Science* Poincaré explained that the antecedents of the crisis in the foundations of science were very old. It had long been sought in vain, he said, to demonstrate the axiom known as Euclid's fifth postulate and this search was the start of the crisis. Euclid's postulate of parallels, which states that through a given point there's not more than one parallel line to a given straight line, we usually learn in tenth-grade geometry. It is one of the basic building blocks out of which the entire mathematics of geometry is constructed.

All the other axioms seemed so obvious as to be unquestionable, but this one did not. Yet you couldn't get rid of it without destroying huge portions of the mathematics, and no one seemed able to reduce it to anything more elementary. What vast effort had been wasted in that chimeric hope was truly unimaginable, Poincaré said.

Finally, in the first quarter of the nineteenth century, and almost at the same time, a Hungarian and a Russian—Bolyai and Lobachevski—established irrefutably that a proof of Euclid's fifth postulate is impossible. They did this by reasoning that if there were any way to reduce Euclid's postulate to other, surer axioms, another effect would also be noticeable: a reversal of Euclid's postulate would create logical contradictions in the geometry. So they reversed Euclid's postulate.

Lobachevski assumes at the start that through a given point can be drawn two parallels to a given straight. And he retains besides all Euclid's other axioms. From these hypotheses he deduces a series of theorems among which it's impossible to find any contradiction, and he constructs a geometry whose faultless logic is inferior in nothing to that of the Euclidian geometry.

Thus by his failure to find any contradiction he proves that the fifth postulate is irreducible to simpler axioms.

It wasn't the proof that was alarming. It was its rational byproduct that soon overshadowed it and almost everything else in the field of mathematics. Mathematics, the cornerstone of scientific certainty, was suddenly uncertain.

We now had *two* contradictory visions of unshakable scientific truth, true for all men of all ages, regardless of their individual preferences.

This was the basis of the profound crisis that shattered the scientific complacency of the Gilded Age. *How do we know which one of these geometries is right?* If there is no basis for distinguishing between them, then you have a total mathematics which admits logical contradictions. But a mathematics that admits internal logical contradictions is no mathematics at all. The ultimate effect of the non-Euclidian geometries becomes nothing more than a magician's mumbo jumbo in which belief is sustained purely by faith!

And of course once that door was opened one could hardly expect the number of contradictory systems of unshakable scientific truth to be limited to two. A German named Riemann appeared with another unshakable system of geometry which throws overboard not only Euclid's postulate, but also the first axiom, which states that only one straight line can pass through two points. Again there is no internal contradiction, only an inconsistency with both Lobachevskian and Euclidian geometries.

According to the Theory of Relativity, Riemann geometry best describes the world we live in.

At Three Forks the road cuts into a narrow canyon of whitish-tan rock, past some Lewis and Clark caves.

East of Butte we go up a long hard grade, cross the Continental Divide, then go down into a valley. Later we pass the great stack of the Anaconda smelter, turn into the town of Anaconda and find a good restaurant with steak and coffee. We go up a long grade that leads to a lake surrounded by pine forests and past some fishermen who push a small boat into the water. Then the road winds down again through the pine forest, and I see by the angle of the sun that the morning is almost ended.

We pass through Phillipsburg and are off into valley meadows. The head wind becomes more gusty here, so I slow down to fifty-five to lessen it a little. We go through Maxville and by the time we reach Hall are badly in need of a rest.

We find a churchyard by the side of the road and stop. The wind is blowing hard now and is chilly, but the sun is warm and we lay out our jackets and helmets on the grass on the leeward side of the church for a rest. It's very lonely and open here, but beautiful. When you have mountains in the distance or even hills, you have space. Chris turns his face into his jacket and tries to sleep.

Everything is so different now without the Sutherlands—so lonely. If you'll excuse me I'll just talk Chautauqua now, until the loneliness goes away.

To solve the problem of what is mathematical truth, Poincaré said, we should first ask ourselves what is the nature of geometric axioms. Are they synthetic *a priori* judgments, as Kant said? That is, do they exist as a fixed part of man's consciousness, independently of experience and uncreated by experience? Poincaré thought not. They would then impose themselves upon us with such force that we couldn't conceive the contrary proposition, or build upon it a theoretic edifice. There would be no non-Euclidian geometry.

Should we therefore conclude that the axioms of geometry are experimental verities? Poincaré didn't think that was so either. If they were, they would be subject to continual change and revision as new laboratory data came in. This seemed to be contrary to the whole nature of geometry itself.

Poincaré concluded that the axioms of geometry are *conventions,* our choice among all possible conventions is *guided* by experimental facts, but it remains *free* and is limited only by the necessity of avoiding all contradiction. Thus it is that the postulates can remain rigorously true even though the experimental laws that have determined their adoption are only approximative. The axioms of geometry, in other words, are merely disguised definitions.

Then, having identified the nature of geometric axioms, he turned to the question, Is Euclidian geometry true or is Riemann geometry true?

He answered, The question has no meaning.

As well ask whether the metric system is true and the avoirdupois system is false; whether Cartesian coordinates are true and polar coordinates are false. One geometry cannot be more true than another; it can only be more *convenient.* Geometry is not true, it is advantageous.

Poincaré then went on to demonstrate the conventional nature of other concepts of science, such as space and time, showing that there isn't one way of measuring these entities that is more true than another; that which is generally adopted is only more *convenient.*

Our concepts of space and time are also definitions, selected on the basis of their convenience in handling the facts.

This radical understanding of our most basic scientific concepts is not yet complete, however. The mystery of what is space and time may be made more understandable by this explanation, but now the burden of sustaining the order of the universe rests on "facts." What are facts?

Poincaré proceeded to examine these critically. *Which* facts are you going to observe? he asked. There is an infinity of them. There is no more chance that an unselective observation of facts will produce science than there is that a monkey at a typewriter will produce the Lord's Prayer.

The same is true of hypotheses. *Which* hypotheses? Poincaré wrote, "If a phenomenon admits of a complete mechanical explanation it will admit of an infinity of others which will account equally well for all the peculiarities disclosed by experiment." This was the statement made by Phaedrus in the laboratory; it raised the question that failed him out of school.

If the scientist had at his disposal infinite time, Poincaré said, it would only be necessary to say to him, "Look and notice well"; but as there isn't time to see everything, and as it's better not to see than to see wrongly, it's necessary for him to make a choice.

Poincaré laid down some rules: There is a hierarchy of facts.

The more general a fact, the more precious it is. Those which serve many times are better than those which have little chance of coming up again. Biologists, for example, would be at a loss to construct a science if only individuals and no species existed, and if heredity didn't make children like parents.

Which facts are likely to reappear? The simple facts. How to recognize them? Choose those that *seem* simple. Either this simplicity is real or the complex elements are indistinguishable. In the first case we're likely to meet this simple fact again either alone or as an element in a complex fact. The second case too has a good chance of recurring since nature doesn't randomly construct such cases.

Where is the simple fact? Scientists have been seeking it in the two extremes, in the infinitely great and in the infinitely small. Biologists, for example, have been instinctively led to regard the cell as more interesting than the whole animal; and, since Poincaré's time, the protein molecule as more interesting than the cell. The outcome has shown the wisdom of this, since cells and molecules belonging to different organisms have been found to be more alike than the organisms themselves.

How then choose the interesting fact, the one that begins again and again? Method is precisely this choice of facts; it is needful then to be occupied first with creating a method; and many have been imagined, since none imposes itself. It's proper to begin with the regular facts, but after a rule is established beyond all doubt, the facts in conformity with it become dull because they no longer teach us anything new. Then it's the exception that becomes important. We seek not resemblances but differences, choose the most accentuated differences because they're the most striking and also the most instructive.

We first seek the cases in which this rule has the greatest chance of failing; by going very far away in space or very far away in time, we may find our usual rules entirely overturned, and these grand over-turnings enable us the better to see the little changes that may happen nearer to us. But what we ought to aim at is less the ascertainment of resemblances and differences than the recognition of likenesses hidden under apparent divergences. Particular rules seem at first discordant, but looking more closely we see in general that they resemble each other; different as to matter, they are alike as to form, as to the order of their parts. When we look at them with this bias we shall see them enlarge and tend to embrace everything. And this it is that makes the value of certain facts that come to complete an assemblage and to show that it is the faithful image of other known assemblages.

No, Poincaré concluded, a scientist does not choose at random the facts he observes. He seeks to condense much experience and much thought into a slender volume; and that's why a little book on physics contains so many past experiences and a thousand times as many possible experiences whose result is known beforehand.

Then Poincaré illustrated how a fact is discovered. He had described generally how scientists arrive at facts and theories but now he penetrated narrowly into his own personal experience with the mathematical functions that established his early fame.

For fifteen days, he said, he strove to prove that there couldn't be any such functions. Every day he seated himself at his work-table, stayed an hour or two, tried a great number of combinations and reached no results.

Then one evening, contrary to his custom, he drank

black coffee and couldn't sleep. Ideas arose in crowds. He felt them collide until pairs interlocked, so to speak, making a stable combination.

The next morning he had only to write out the results. A wave of crystallization had taken place.

He described how a second wave of crystallization, guided by analogies to established mathematics, produced what he later named the "Theta-Fuchsian Series." He left Caen, where he was living, to go on a geologic excursion. The changes of travel made him forget mathematics. He was about to enter a bus, and at the moment when he put his foot on the step, the idea came to him, without anything in his former thoughts having paved the way for it, that the transformations he had used to define the Fuchsian functions were identical with those of non-Euclidian geometry. He didn't verify the idea, he said, he just went on with a conversation on the bus; but he felt a perfect certainty. Later he verified the result at his leisure.

A later discovery occurred while he was walking by a seaside bluff. It came to him with just the same characteristics of brevity, suddenness and immediate certainty. Another major discovery occurred while he was walking down a street. Others eulogized this process as the mysterious workings of genius, but Poincaré was not content with such a shallow explanation. He tried to fathom more deeply what had happened.

Mathematics, he said, isn't merely a question of applying rules, any more than science. It doesn't merely make the most combinations possible according to certain fixed laws. The combinations so obtained would be exceedingly numerous, useless and cumbersome. The true work of the inventor consists in choosing among these combinations so as to eliminate the useless ones, or rather, to avoid the trouble of making them, and the rules that must guide the choice are extremely fine and delicate. It's almost impossible to state them precisely; they must be felt rather than formulated.

Poincaré then hypothesized that this selection is made by what he called the "subliminal self," an entity that corresponds exactly with what Phaedrus called preintellectual awareness. The subliminal self, Poincaré said, looks at a large number of solutions to a problem, but only the *interesting* ones break into the domain of consciousness. Mathematical solutions are selected by the subliminal self on the basis of "mathematical beauty," of the harmony of numbers and forms, of geometric elegance. "This is a true esthetic feeling which all mathematicians know," Poincaré said, "but of which the profane are so ignorant as often to be tempted to smile." But it is this harmony, this beauty, that is at the center of it all.

Poincaré made it clear that he was not speaking of romantic beauty, the beauty of appearances which strikes the senses. He meant classic beauty, which comes from the harmonious order of the parts, and which a pure intelligence can grasp, which gives structure to romantic beauty and without which life would be only vague and fleeting, a dream from which one could not distinguish one's dreams because there would be no basis for making the distinction. It is the quest of this special classic beauty, the sense of harmony of the cosmos, which makes us *choose the facts most fitting to contribute to this harmony*. It is not the facts but the relation of things that results in the universal harmony that is the sole objective reality.

What guarantees the objectivity of the world in which we live is that this world is common to us with other thinking beings. Through the communications that we have with other men we receive from them ready-made harmonious reasonings. We know that these reasonings do not come from us and at the same time we recognize in them, *because of their harmony,* the work of reasonable beings like ourselves. And as these reasonings appear to fit the world of our sensations, we think we may infer that these reasonable beings have seen the same thing as we; thus it is that we know we haven't been dreaming. It is this harmony, this *quality* if you will, that is the sole basis for the only reality we can ever know.

Poincaré's contemporaries refused to acknowledge that facts are preselected because they thought that to do so would destroy the validity of scientific method. They presumed that "preselected facts" meant that truth is "whatever you like" and called his ideas conventionalism. They vigorously ignored the truth that their own "principle of objectivity" is not itself an observable fact—and therefore by their own criteria should be put in a state of suspended animation.

They felt they had to do this because if they didn't, the entire philosophic underpinning of science would collapse. Poincaré didn't offer any resolutions of this quandary. He didn't go far enough into the metaphysical implications of what he was saying to arrive at the solution. What he neglected to say was that the selection of facts before you "observe" them is "whatever you like" *only in a dualistic, subject-object metaphysical system!* When Quality enters the picture as a third metaphysical entity, the preselection of facts is no longer arbitrary. The preselection of facts is not based on subjective, capricious "whatever you like" but on *Quality,* which is reality itself. Thus the quandary vanishes.

It was as though Phaedrus had been working on a puzzle of his own and because of lack of time had left one whole side unfinished.

Poincaré had been working on a puzzle of *his* own. His judgment that the scientist selects facts, hypotheses and axioms on the basis of harmony also left the rough serrated edge of a puzzle incomplete. To leave the impression in the scientific world that the source of all scientific reality is merely a subjective, capricious harmony is to solve problems of epistemology while leaving an unfinished edge at the border of metaphysics that makes the epistemology unacceptable.

But we know from Phaedrus' metaphysics that the harmony Poincaré talked about is *not subjective.* It is the *source* of subjects and objects and exists in an anterior relationship to them. It is *not* capricious, it is the force that *opposes* capriciousness; the ordering principle of all scientific and mathematical thought which *destroys* capriciousness, and without which no scientific thought can proceed. What brought tears of recognition to my eyes was the discovery that these unfinished edges match perfectly in a kind of harmony that both Phaedrus and Poincaré talked about, to produce a complete structure of thought capable of uniting the separate languages of Science and Art into one.

On either side of us the mountains have become steep, to form a long narrow valley that winds into Missoula. This head wind has worn me down and I'm tired now.

Chris taps me and points to a high hill with a large painted *M* on it. I nod. This morning we passed one like it as we left Bozeman. A fragment occurs to me that the freshmen in each school go up there and paint the *M* each year.

At a station where we fill with gas, a man with a trailer carrying two Appaloosa horses strikes up a conversation. Most horse people are antimotorcycle, it seems, but this one is not, and he asks a lot of questions, which I answer. Chris keeps asking to go up to the *M,* but I can see from here it's a steep, rutty, scrambler road. With our highway machine and heavy load I don't want to fool with it. We stretch our legs for a while, walk around and then somewhat wearily head out of Missoula toward Lolo Pass.

A recollection appears that not many years ago this road was all dirt with twists and turns around every rock and fold in the mountains. Now it's paved and the turns are very broad. All the traffic we were in has evidently headed north for Kalispell or Coeur D'Alene, for there's hardly any now. We're headed southwest, have picked up a tail wind, and we feel better because of it. The road starts to wind up into the pass.

All traces of the East are gone now, at least in my imagination. All the rain here comes from Pacific winds and all the rivers and streams here return it to the Pacific. We should be at the ocean in two or three days.

At Lolo Pass we see a restaurant, and pull up in front of it beside an old Harley high-miler. It has a homemade pannier on the back and thirty-six thousand on the odometer. A real cross-country man.

Inside we fill up on pizza and milk, and when finished leave right away. There's not much sunlight left, and a search for a campsite after dark is difficult and unpleasant.

As we leave we see the cross-country man by the cycles with his wife and we say hello. He is from Missouri, and the relaxed look on his wife's face tells me they've been having a good trip.

The man asks, "Were you bucking that wind up to Missoula too?"

I nod. "It must have been thirty or forty miles an hour."

"At least," he says.

We talk about camping for a while and they comment on how cold it is. They never dreamed in Missouri it would be this cold in the summer, even in the mountains. They've had to buy clothes and blankets.

"It shouldn't be too cold tonight," I say. "We're only at about five thousand feet."

Chris says, "We're going to camp just down the road."

"At one of the campsites?"

"No, just somewhere off the road," I say.

They show no inclination of wanting to join us, so after a pause I press the starter button and we wave off.

On the road the shadows of the mountain trees are long now. After five or ten miles we see some logging road turnoffs and head up.

The logging road is sandy, so I keep in low gear with feet out to prevent a spill. We see side roads off the main logging road but I stay on the main one until after about a mile we come to some bulldozers. That means they're still logging here. We turn back and head up one of the side roads. After about half a mile we come to a tree fallen across the road. That's good. That means this road has been abandoned.

I say, "This is it" to Chris, and he gets off. We're on a slope that allows us to see over unbroken forest for miles.

Chris is all for exploring, but I'm so tired I just want to rest. "You go by yourself," I say.

"No, you come along."

"I'm really tired, Chris. In the morning we'll explore."

I untie the packs and spread the sleeping bags out on the ground. Chris goes off. I stretch out, and the tiredness fills my arms and legs. Silent, beautiful forest. . . .

In time Chris returns, and says he has diarrhea.

"Oh," I say, and get up. "Do you have to change underwear?"

"Yes." He looks sheepish.

"Well, they're in the pack by the front of the cycle. Change and get a bar of soap from the saddlebag and we'll go down to the stream and wash the old underwear out." He's embarrassed by the whole thing and now is glad to take orders.

The downward slope of the road makes our feet flop as we head toward the stream. Chris shows me some stones he's collected while I've been sleeping. The pine smell of the forest is rich here. It's turning cool and the sun is very low. The silence and the fatigue and the sinking of the sun depress me a little, but I keep it to myself.

After Chris has washed out his underwear and has it completely clean and wrung out we head back up the logging road. As we climb it I get a sudden depressed feeling I've been walking up this logging road all my life.

"Dad?"

"What?" A small bird rises from a tree in front of us.

"What should I be when I grow up?"

The bird disappears over a far ridge. I don't know what to say. "Honest," I finally say.

"I mean what kind of a job?"

"Any kind."

"Why do you get mad when I ask that?"

"I'm not mad . . . I just think . . . I don't know . . . I'm just too tired to think. . . . It doesn't matter what you do."

Roads like this one get smaller and smaller and then quit.

Later I notice he's not keeping up.

The sun is below the horizon now and twilight is on us. We walk separately back up the logging road and when we reach the cycle we climb into the sleeping bags and without a word go to sleep.

23

There it is at the end of the corridor: a glass door. And behind it are Chris and on one side of him his younger brother and on the other side his mother. Chris has his hand against the glass. He recognizes me and waves. I wave back and approach the door.

How silent everything is. Like watching a motion picture when the sound has failed.

Chris looks up at his mother and smiles. She smiles down at him but I see she is only covering her grief. She's very distressed about something but she doesn't want them to see.

And now I see what the glass door is. It is the door of a coffin—mine.

Not a coffin, a sarcophagus. I am in an enormous vault, dead, and they are paying their last respects.

It's kind of them to come and do this. They didn't have to do this. I feel grateful.

Now Chris motions for me to open the glass door of the vault. I see he wants to talk to me. He wants me to tell him, perhaps, what death is like. I feel a desire to do this, to tell him. It was so good of him to come and wave I will tell him it's not so bad. It's just lonely.

I reach to push the door open but a dark figure in a shadow beside the door motions for me not to touch it. A single finger is raised to lips I cannot see. The dead aren't permitted to speak.

But they *want* me to talk. I'm still *needed!* Doesn't he see this? There must be some kind of mistake. Doesn't he see that they need me? I plead with the figure that I have to speak to them. It's not finished yet. I have to tell them things. But the figure in the shadows makes no sign he has even heard.

"Chris!" I shout through the door. *"I'll see you!!"* The dark figure moves toward me threateningly, but I hear Chris's voice, "Where?" faint and distant. He *heard* me! And the dark figure, enraged, draws a curtain over the door.

Not the mountain, I think. The mountain is gone. *"At the bottom of the ocean!!"* I shout.

And now I am standing in the deserted ruins of a city all alone. The ruins are all around me endlessly in every direction and I must walk them alone.

24

The sun is up.

For a while I'm not sure where I am.

We're on a road in a forest somewhere.

Bad dream. That glass door again.

The chrome of the cycle gleams beside me and then I see the pines and then Idaho comes to mind.

The door and the shadowy figure beside it were just imaginary.

We're on a logging road, that's right . . . bright day . . . sparkling air. Wow! . . . it's beautiful. We're headed for the ocean.

I remember the dream again and the words "I'll see you at the bottom of the ocean" and wonder about them. But pines and sunlight are stronger than any dream and the wondering goes away. Good old reality.

I get out of the sleeping bag. It's cold and I get dressed quickly. Chris is asleep. I walk around him, climb over a fallen tree trunk and walk up the logging road. To warm myself I speed up to a jog and move up the road briskly. Good, good, good, good, good. The word keeps time with the jogging. Some birds fly up from the shadowy hill into the sunlight and I watch them until they're out of sight. Good, good, good, good, good. Crunchy gravel on the road. Good, good. Bright yellow sand in the sun. Good, good, good. These roads go on for miles sometimes. Good, good, good.

Eventually I reach a point where I'm really winded. The road is higher now and I can see for miles over the forest.

Good.

Still puffing, I walk back down at a brisk pace, crunching more gently now, noticing small plants and shrubs where the pines have been logged.

At the cycle again I pack gently and quickly. By now I'm so familiar with how everything goes together it's almost done without thought. Finally I need Chris's sleeping bag. I roll him a little, not too rough, and tell him, "Great day!"

He looks around, disoriented. He gets out of the sleeping bag and, while I pack it, gets dressed without really knowing what he does.

"Put your sweater and jacket on," I say. "It's going to be a chilly ride."

He does and gets on and in low gear we follow the logging road down to where it meets the blacktop again. Before we start on it I take one last look back up. Nice. A nice spot. From here the blacktop winds down and down.

Long Chautauqua today. One that I've been looking forward to during the whole trip.

Second gear and then third. Not too fast on these curves. Beautiful sunlight on these forests.

There has been a haze, a backup problem in this Chautauqua so far; I talked about caring the first day and then realized I couldn't say anything meaningful about caring until its inverse side, Quality, is understood. I think it's important now to tie care to Quality by pointing out that care and Quality are internal and external aspects of the same thing. A person who sees Quality and feels it as he works is a person who cares. A person who cares about what he sees and does is a person who's bound to have some characteristics of Quality.

Thus, if the problem of technological hopelessness is caused by absence of care, both by technologists and antitechnologists; and if care and Quality are external and internal aspects of the same thing, then it follows logically that what really causes technological hopelessness is absence of the perception of Quality in technology by both technologists and antitechnologists. Phaedrus' mad pursuit of the rational, analytic and therefore *technological* meaning of the word "Quality" was really a pursuit of the answer to the whole problem of technological hopelessness. So it seems to me, anyway.

So I backed up and shifted to the classic-romantic split that I think underlies the whole humanist-technological problem. But that too required a backup into the meaning of Quality.

But to understand the meaning of Quality in classic terms required a backup into metaphysics and its relationship to everyday life. To do that required still another backup into the huge area that relates both metaphysics and everyday life—namely, formal reason. So I proceeded with formal reason up into metaphysics and then into Quality and then from Quality back down into metaphysics and science.

Now we go still further down from science into technology, and I do believe that at last we are where I wanted to be in the first place.

But now we have with us some concepts that greatly alter the whole understanding of things. Quality is the Buddha. Quality is scientific reality. Quality is the goal of Art. It remains to work these concepts into a practical, down-to-earth context, and for this there is nothing more practical or down-to-earth than what I have been talking about all along—the repair of an old motorcycle.

This road keeps on winding down through this canyon. Early morning patches of sun are around us everywhere. The cycle hums through the cold air and mountain pines and we pass a small sign that says a breakfast place is a mile ahead.

"Are you hungry?" I shout.

"Yes!" Chris shouts back.

Soon a second sign saying CABINS with an arrow under it points off to the left. We slow down, turn and follow a dirt road until it reaches some varnished log cabins under some trees. We pull the cycle under a tree, shut off the ignition and gas and walk inside the main lodge. The wooden floors have a nice clomp under the cycle boots. We sit down at a tableclothed table and order eggs, hot cakes, maple syrup, milk, sausages and orange juice. That cold wind has worked up an appetite.

oneers, but which, like "kin," seems to have all but dropped out of use. I like it also because it describes exactly what happens to someone who connects with Quality. He gets filled with gumption.

The Greeks called it *enthousiasmos,* the root of "enthusiasm," which means literally "filled with *theos,*" or God, or Quality. See how that fits?

A person filled with gumption doesn't sit around dissipating and stewing about things. He's at the front of the train of his own awareness, watching to see what's up the track and meeting it when it comes. That's gumption.

Chris arrives and says, "I'm feeling better now."

"Good," I say. We pack up the soap and toilet paper and put the towel and wet underwear where they won't get other things damp and then we get on and are moving again.

The gumption-filling process occurs when one is quiet long enough to see and hear and feel the real universe, not just one's own stale opinions about it. But it's nothing exotic. That's why I like the word.

You see it often in people who return from long, quiet fishing trips. Often they're a little defensive about having put so much time to "no account" because there's no intellectual justification for what they've been doing. But the returned fisherman usually has a peculiar abundance of gumption, usually for the very same things he was sick to death of a few weeks before. He hasn't been wasting time. It's only our limited cultural viewpoint that makes it seem so.

If you're going to repair a motorcycle, an adequate

supply of gumption is the first and most important tool. If you haven't got that you might as well gather up all the other tools and put them away, because they won't do you any good.

Gumption is the psychic gasoline that keeps the whole thing going. If you haven't got it there's no way the motorcycle can possibly be fixed. But if you *have* got it and know how to keep it there's absolutely no way in this whole world that motorcycle can *keep* from getting fixed. It's bound to happen. Therefore the thing that must be monitored at all times and preserved before anything else is the gumption.

This paramount importance of gumption solves a problem of format of this Chautauqua. The problem has been how to get off the generalities. If the Chautauqua gets into the actual details of fixing one individual machine the chances are overwhelming that it won't be your make and model and the information will be not only useless but dangerous, since information that fixes one model can sometimes wreck another. For detailed information of an objective sort, a separate shop manual for the specific make and model of machine must be used. In addition, a general shop manual such as *Audel's Automotive Guide* fills in the gaps.

But there's another kind of detail that no shop manual goes into but that is common to all machines and can be given here. This is the detail of the Quality relationship, the gumption relationship, between the machine and the mechanic, which is just as intricate as the machine itself. Throughout the process of fixing the machine things always come up, low-quality things, from a dusted knuckle to an accidentally ruined "irre-

"I want to write a letter to Mom," Chris says.

That sounds good to me. I go to the desk and get some of the lodge stationery. I bring it to Chris and give him my pen. That brisk morning air has given him some energy too. He puts the paper in front of him, grabs the pen in a heavy grip and then concentrates on the blank paper for a while.

He looks up. "What day is it?"

I tell him. He nods and writes it down.

Then I see him write, "Dear Mom:"

Then he stares at the paper for a while.

Then he looks up. "What should I say?"

I start to grin. I should have him write for an hour about one side of a coin. I've sometimes thought of him as a student but not as a rhetoric student.

We're interrupted by the hot cakes and I tell him to put the letter to one side and I'll help him afterward.

When we are done I sit smoking with a leaden feeling from the hot cakes and the eggs and everything and notice through the window that under the pines outside the ground is in patches of shadow and sunlight.

Chris brings out the paper again. "Now help me," he says.

"Okay," I say. I tell him getting stuck is the commonest trouble of all. Usually, I say, your mind gets stuck when you're trying to do too many things at once. What you have to do is try not to force words to come. That just gets you more stuck. What you have to do now is separate out the things and do them one at a time. You're trying to think of what to *say* and what to say *first* at the same time and that's too hard. So separate them out. Just make a list of all the things you want to say in any old order. Then later we'll figure out the right order.

"Like what things?" he asks.

"Well, what do you want to tell her?"

"About the trip."

"What things about the trip?"

He thinks for a while. "About the mountain we climbed."

"Okay, write that down," I say.

He does.

Then I see him write down another item, then another, while I finish my cigarette and coffee. He goes through three sheets of paper, listing things he wants to say.

"Save those," I tell him, "and we'll work on them later."

"I'll never get all this into one letter," he says.

He sees me laugh and frowns.

I say, "Just pick out the best things." Then we head outside and onto the motorcycle again.

On the road down the canyon now we feel the steady drop of altitude by a popping of ears. It's becoming warmer and the air is thicker too. It's good-bye to the high country, which we've been more or less in since Miles City.

Stuckness. That's what I want to talk about today.

Back on our trip out of Miles City you'll remember I talked about how formal scientific method could be applied to the repair of a motorcycle through the study of chains of cause and effect and the application of experimental method to determine these chains. The purpose then was to show what was meant by classic rationality.

Now I want to show that that classic pattern of rationality can be tremendously improved, expanded and made far more effective through the formal recognition of Quality in its operation. Before doing this, however, I should go over some of the negative aspects of traditional maintenance to show just where the problems are.

The first is stuckness, a mental stuckness that accompanies the physical stuckness of whatever it is you're working on. The same thing Chris was suffering from. A screw sticks, for example, on a side cover assembly. You check the manual to see if there might be any special cause for this screw to come off so hard, but all it says is "Remove side cover plate" in that wonderful terse technical style that never tells you what you want to know. There's no earlier procedure left undone that might cause the cover screws to stick.

If you're experienced you'd probably apply a penetrating liquid and an impact driver at this point. But suppose you're inexperienced and you attach a self-locking plier wrench to the shank of your screwdriver and really twist it hard, a procedure you've had success with in the past, but which this time succeeds only in tearing the slot of the screw.

Your mind was already thinking ahead to what you would do when the cover plate was off, and so it takes a little time to realize that this irritating minor annoyance of a torn screw slot isn't just irritating and minor. You're stuck. Stopped. Terminated. It's absolutely stopped you from fixing the motorcycle.

This isn't a rare scene in science or technology. This is the commonest scene of all. Just plain *stuck*. In traditional maintenance this is the worst of all moments, so bad that you have avoided even thinking about it before you come to it.

The book's no good to you now. Neither is scientific reason. You don't need any scientific experiments to find out what's wrong. It's obvious what's wrong. What you need is an hypothesis for how you're going to get that slotless screw out of there and scientific method doesn't provide any of these hypotheses. It operates only after they're around.

This is the zero moment of consciousness. Stuck. No answer. Honked. Kaput. It's a miserable experience emotionally. You're losing time. You're incompetent. You don't know what you're doing. You should be ashamed of yourself. You should take the machine to a *real* mechanic who knows how to figure these things out.

It's normal at this point for the fear-anger syndrome to take over and make you want to hammer on that side plate with a chisel, to pound it off with a sledge if necessary. You think about it, and the more you think about it the more you're inclined to take the whole machine to a high bridge and drop it off. It's just outrageous that a tiny little slot of a screw can defeat you so totally.

What you're up against is the great unknown, the void of all Western thought. You need some ideas, some hypotheses. Traditional scientific method, unfortunately, has never quite gotten around to say exactly where to pick up more of these hypotheses. Traditional scientific method has always been, at the very *best,* 20-20 hindsight. It's good for seeing where you've been. It's good for testing the truth of what you think you know, but it can't tell you where you *ought* to go, unless where you ought to go is a continuation of where you were going in the past. Creativity, originality, inventiveness, intuition, imagination—"unstuckness," in other words—are completely outside its domain.

We continue down the canyon, past folds in the steep slopes where wide streams enter. We notice the river grows rapidly now as streams enlarge it. Turns in the road are less sharp here and straight stretches are longer. I move into the highest gear.

Later the trees become scarce and spindly, with large areas of grass and underbrush between them. It's too hot for the jacket and sweater so I stop at a roadside pulloff to remove them.

Chris wants to go hiking up a trail and I let him, finding a small shady spot to sit back and rest. Just quiet now, and meditative.

A display describes a fire burn that took place here years ago. According to the information the forest is filling in again but it will be years before it returns to its former condition.

Later the crunch of gravel tells me Chris is coming back down the trail. He didn't go very far. When he arrives he says, "Let's go." We retie the pack, which has started to shift a little, and then move out on the highway. The sweat from sitting there cools suddenly from the wind.

We're still stuck on that screw and the only way it's going to get unstuck is by abandoning further examination of the screw according to traditional scientific method. That won't work. What we have to do is examine traditional scientific method in the light of that stuck screw.

We have been looking at that screw "objectively." According to the doctrine of "objectivity," which is integral with traditional scientific method, what we like or don't like about that screw has nothing to do with our correct thinking. We should not evaluate what we see. We should keep our mind a blank tablet which nature fills for us, and then reason disinterestedly from the facts we observe.

But when we stop and think about it disinterestedly, in terms of this stuck screw, we begin to see that this whole idea of disinterested observation is silly. Where *are* those facts? What are we going to observe disinterestedly? The torn slot? The immovable side cover plate? The color of the paint job? The speedometer? The sissy bar? As Poincaré would have said, there are an infinite number of facts about the motorcycle, and the right ones don't just dance up and introduce themselves. The right facts, the ones we really need, are not only passive, they are damned *elusive,* and we're not going to just sit back and "observe" them. We're going to have to be in there *looking* for them or we're going to be here a long time. Forever. As Poincaré pointed out, there *must* be a subliminal choice of what facts we observe.

The difference between a good mechanic and a bad one, like the difference between a good mathematician and a bad one, is precisely this ability to *select* the good facts from the bad ones on the basis of quality. He has to *care*! This is an ability about which formal traditional scientific method has nothing to say. It's long past time to take a closer look at this qualitative preselection of facts which has seemed so scrupulously ignored by those who make so much of these facts after they are "observed." I think that it will be found that a formal acknowledgment of the role of Quality in the scientific process doesn't destroy the empirical vision at all. It expands it, strengthens it and brings it far closer to actual scientific practice.

I think the basic fault that underlies the problem of stuckness is traditional rationality's insistence upon "objectivity," a doctrine that there is a divided reality of subject and object. For true science to take place these must be rigidly separate from each other. "You are the mechanic. There is the motorcycle. You are forever apart from one another. You do this to it. You do that to it. These will be the results."

This eternally dualistic subject-object way of approaching the motorcycle sounds right to us because we're used to it. But it's not right. It's always been an artificial interpretation *superimposed* on reality. It's never been reality itself. When this duality is completely accepted a certain nondivided relationship between the mechanic and motorcycle, a craftsmanlife feeling for the work, is destroyed. When traditional rationality divides the world into subjects and objects it shuts out Quality, and when you're really stuck it's Quality, not any subjects or objects, that tells you where you ought to go.

By returning our attention to Quality it is hoped that we can get technological work out of the noncaring subject-object dualism and back into craftsmanlike self-involved reality again, which will reveal to us the facts we need when we are stuck.

In my mind now is an image of a huge, long railroad

train, one of those 120-boxcar jobs that cross the prairies all the time with lumber and vegetables going east and with automobiles and other manufactured goods going west. I want to call this railroad train "knowledge" and subdivide it into two parts: Classic Knowledge and Romantic Knowledge.

In terms of the analogy, Classic Knowledge, the knowledge taught by the Church of Reason, is the engine and all the boxcars. All of them and everything that's in them. If you subdivide the train into parts you will find no Romantic Knowledge anywhere. And unless you're careful it's easy to make the presumption that's all the train there is. This isn't because Romantic Knowledge is nonexistent or even unimportant. It's just that so far the definition of the train is static and purposeless. This was what I was trying to get at back in South Dakota when I talked about two whole dimensions of existence. It's two whole ways of *looking* at the train.

Romantic Quality, in terms of this analogy, isn't any "part" of the train. It's the leading edge of the engine, a two-dimensional surface of no real significance unless you understand that the train isn't a static entity at all. A train really isn't a train if it can't go anywhere. In the process of examining the train and subdividing it into parts we've inadvertently stopped it, so that it really isn't a train we are examining. That's why we get stuck.

The real train of knowledge isn't a static entity that can be stopped and subdivided. It's always going somewhere. On a track called Quality. And that engine and all those 120 boxcars are never going anywhere except where the track of Quality takes them; and romantic Quality, the leading edge of the engine, takes them along that track.

Romantic reality is the cutting edge of experience. It's the leading edge of the train of knowledge that keeps the whole train on the track. Traditional knowledge is only the collective memory of where that leading edge has been. At the leading edge there are no subjects, no objects, only the track of Quality ahead, and if you have no formal way of evaluating, no way of acknowledging this Quality, then the entire train has no way of knowing where to go. You don't have pure reason—you have pure confusion. The leading edge is where absolutely all the action is. The leading edge contains all the infinite possibilities of the future. It contains all the history of the past. Where else could they be contained?

The past cannot remember the past. The future can't generate the future. The cutting edge of this instant right here and now is always nothing less than the totality of everything there is.

Value, the leading edge of reality, is no longer an irrelevant offshoot of structure. Value is the predecessor of structure. It's the preintellectual awareness that gives rise to it. Our structured reality is preselected on the basis of value, and really to understand structured reality requires an understanding of the value source from which it's derived.

One's rational understanding of a motorcycle is therefore modified from minute to minute as one works on it and sees that a new and different rational understanding has more Quality. One doesn't cling to old sticky ideas because one has an immediate rational basis for rejecting them. Reality isn't static anymore.

It's not a set of ideas you have to either fight or resign yourself to. It's made up, in part, of ideas that are expected to grow as you grow, and as we all grow, century after century. With Quality as a central undefined term, reality is, in its essential nature, not static but dynamic. And when you really understand dynamic reality you never get stuck. It has forms but the forms are capable of change.

To put it in more concrete terms: If you want to build a factory, or fix a motorcycle, or set a nation right without getting stuck, then classical, structured, dualistic subject-object knowledge, although necessary, isn't enough. You have to have some feeling for the quality of the work. You have to have a sense of what's good. *That* is what carries you forward. This sense isn't just something you're born with, although you *are* born with it. It's also something you can develop. It's not just "intuition," not just unexplainable "skill" or "talent." It's the direct result of contact with basic *reality,* Quality, which dualistic reason has in the past tended to conceal.

It all sounds so far out and esoteric when it's put like that it comes as a shock to discover that it is one of the most homespun, down-to-earth views of reality you can have. Harry Truman, of all people, comes to mind, when he said, concerning his administration's programs, "We'll just try them . . . and if they don't work . . . why then we'll just try something else." That may not be an exact quote, but it's close.

The reality of the American government isn't static, he said, it's *dynamic.* If we don't like it we'll get something better. The American government isn't going to get stuck on any set of fancy doctrinaire ideas.

The key word is "better"—Quality. Some may argue that the underlying form of the American government *is* stuck, *is* incapable of change in response to Quality, but that argument is not to the point. The point is that the President and everyone else, from the wildest radical to the wildest reactionary, agree that the government *should* change in response to Quality, even if it doesn't. Phaedrus' concept of changing Quality as reality, a reality so ominpotent that whole governments must change to keep up with it, is something that in a wordless way we have always unanimously believed in all along.

And what Harry Truman said, really, was nothing different from the practical, pragmatic attitude of any laboratory scientist or any engineer or any mechanic when he's not thinking "objectively" in the course of his daily work.

I keep talking wild theory, but it keeps somehow coming out stuff everybody knows, folklore. This Quality, this feeling for the work, is something known in every shop.

Now finally let's get back to that screw.

Let's consider a reevaluation of the situation in which we assume that the stuckness now occurring, the zero of consciousness, isn't the worst of all possible situations, but the best possible situation you could be in. After all, it's exactly this stuckness that Zen Buddhists go to so much trouble to induce; through koans, deep breathing, sitting still and the like. Your mind is empty, you have a "hollow-flexible" attitude of "beginner's mind." You're right at the front end of the train of knowledge, at the track of reality itself. Consider, for a change, that this is a moment to be not feared but cultivated. If your mind is truly, profoundly stuck, then you may be much better off than when it was loaded with ideas.

The solution to the problem often at first seems unimportant or undesirable, but the state of stuckness allows it, in time, to assume its true importance. It seemed small because your previous rigid evaluation which led to the stuckness made it small.

But now consider the fact that no matter how hard you try to hang on to it, this stuckness is bound to disappear. Your mind will naturally and freely move toward a solution. Unless you are a real master at staying stuck you can't prevent this. The fear of stuckness is needless because the longer you stay stuck the more you see the Quality-reality that gets you unstuck every time. What's *really* been getting you stuck is the running from the stuckness through the cars of your train of knowledge looking for a solution that is out in front of the train.

Stuckness shouldn't be avoided. It's the psychic predecessor of all real understanding. An egoless acceptance of stuckness is a key to an understanding of all Quality, in mechanical work as in other endeavors. It's this understanding of Quality as revealed by stuckness which so often makes self-taught mechanics so superior to institute-trained men who have learned how to handle everything except a new situation.

Normally screws are so cheap and small and simple you think of them as unimportant. But now, as your Quality awareness becomes stronger, you realize that this one, individual, particular screw is neither cheap nor small nor unimportant. Right now this screw is worth exactly the selling price of the whole motorcycle, because the motorcycle is actually valueless until you get the screw out. With this reevaluation of the screw comes a willingness to expand your knowledge of it.

With the expansion of the knowledge, I would guess, would come a reevaluation of what the screw really is. If you concentrate on it, think about it, stay stuck on it for a long enough time, I would guess that in time you will come to see that the screw is less and less an object typical of a class and more an object unique in itself. Then with more concentration you will begin to see the screw as not even an object at all but as a collection of functions. Your stuckness is gradually eliminating patterns of traditional reason.

In the past when you separated subject and object from one another in a permanent way, your thinking about them got very rigid. You formed a class called "screw" that seemed to be inviolable and more real than the reality you are looking at. And you couldn't think of how to get unstuck because you couldn't think of anything new, because you couldn't *see* anything new.

Now, in getting that screw out, you aren't interested in what it *is*. What it *is* has ceased to be a category of thought and is a continuing direct experience. It's not in the boxcars anymore, it's out in front and capable of change. You are interested in what it *does* and why it's doing it. You will ask functional questions. Associated with your questions will be a subliminal Quality discrimination identical to the Quality discrimination that led Poincaré to the Fuchsian equations.

What your actual solution is is unimportant as long as it has Quality. Thoughts about the screw as combined rigidness and adhesiveness and about its special helical interlock might lead naturally to solutions of impaction and use of solvents. That is one kind of Quality track. Another track may be to go to the library and look through a catalog of mechanic's tools, in which you might come across a screw extractor that would do the job. Or to call a friend who knows something about mechanical work. Or just to drill the screw out, or just burn it out with a torch. Or you might just, as a result of your meditative attention to the screw, come up with some new way of extracting it that has never been thought of before and that beats all the rest and is patentable and makes you a millionaire five years from now. There's no predicting what's on that Quality track. The solutions all are simple—after you have arrived at them. But they're simple only when you know already what they are.

Highway 13 follows another branch of our river but now it goes upstream past old sawmill towns and sleepy scenery. Sometimes when you switch from a federal to a state highway it seems like you drop back like this in time. Pretty mountains, pretty river, bumpy but pleasant tar road . . . old buildings, old people on a front porch . . . strange how old, obsolete buildings and plants and mills, the technology of fifty and a hundred years ago, always seem to look so much better than the new stuff. Weeds and grass and wildflowers grow where the concrete has cracked and broken. Neat, squared, upright lines acquire a random sag. The uniform masses of the unbroken color of fresh paint modify to a mottled, weathered softness. Nature has a non-Euclidian geometry of her own that seems to soften the deliberate objectivity of these buildings with a kind of random spontaneity that architects would do well to study.

Soon we leave the river and the old sleepy buildings and now climb to some sort of a dry, meadowy plateau. The road rolls and bumps and rocks so much I have to keep the speed down to fifty. There are some bad chuckholes in the asphalt and I watch carefully for more.

We're really accustomed to making mileage. Stretches that would have seemed long back in the Dakotas now seem short and easy. Being on the machine seems more natural than being off it. We're nowhere that I'm familiar with, in country that I've never seen before, yet I don't feel a stranger in it.

At the top of the plateau at Grangeville, Idaho, we step from the blasting heat into an air-conditioned restaurant. Deep cool inside. While we wait for chocolate malteds I notice a high-schooler sitting at the counter exchanging looks with the girl next to him. She's gorgeous, and I'm not the only other one who notices it. The girl behind the counter waiting on them is also watching with an anger she thinks no one else sees. Some kind of triangle. We keep passing unseen through little moments of other people's lives.

Back in the heat again and not far from Grangeville we see that the dry plateau that looked almost like prairie when we were out on it suddenly breaks away into an enormous canyon. I see our road will go down and down through what must be a hundred hairpin turns into a desert of broken land and crags. I tap Chris's knee and point and as we round a turn where we see it all I hear him holler, "Wow!"

At the brink I shift down to third, then close the throttle. The engine drags, backfiring a little, and down we go.

By the time our cycle has reached the bottom of wherever it is we are, we have dropped thousands of feet. I look back over my shoulder and see antlike cars way back at the top. Now we must head forward across this baking desert to wherever the road leads.

25

This morning a solution to the problem of stuckness was discussed, the classic badness caused by traditional reason. Now it's time to move to its romantic parallel, the ugliness of the technology traditional reason has produced.

The road has twisted and rolled over desert hills into a little, narrow thread of green surrounding the town of White Bird, then proceeded on to a big fast river, the Salmon, flowing between high canyon walls. Here the heat is tremendous and the glare from the white canyon rock is blinding. We wind on and on through the bottom of the narrow canyon, nervous about fast-moving traffic and oppressed by the fiery heat.

The ugliness the Sutherlands were fleeing is not inherent in technology. It only seemed that way to them because it's so hard to isolate what it is within technology that's so ugly. But technology is simply the making of things and the making of things can't by its own nature be ugly or there would be no possibility for beauty in the arts, which also include the making of things. Actually a root word of technology, *techne,* originally *meant* "art." The ancient Greeks never separated art from manufacture in their minds, and so never developed separate words for them.

Neither is the ugliness inherent in the materials of modern technology—a statement you sometimes hear. Mass-produced plastics and synthetics aren't in themselves bad. They've just acquired bad associations. A person who's lived inside stone walls of a prison most of his life is likely to see stone as an inherently ugly material, even though it's also the prime material of sculpture, and a person who's lived in a prison of ugly plastic technology that started with his childhood toys and continues through a lifetime of junky consumer products is likely to see this material as inherently ugly. But the real ugliness of modern technology isn't found in any material or shape or act or product. These are just the objects in which the low Quality appears to reside. It's our habit of assigning Quality to subjects or objects that gives this impression.

The real ugliness is not the result of any objects of technology. Nor is it, if one follows Phaedrus' metaphysics, the result of any subjects of technology, the people who produce it or the people who use it. Quality, or its absence, doesn't reside in either the subject or the object. The real ugliness lies in the relationship between the people who produce the technology and the things they produce, which results in a similar relationship between the people who use the technology and the things they use.

Phaedrus felt that at the moment of pure Quality perception, or not even perception, at the moment of pure Quality, there is no subject and there is no object. There is only a sense of Quality that produces a later awareness of subjects and objects. At the moment of pure quality, subject and object are identical. This is the *Tat tvam asi* truth of the Upanishads, but it's also reflected in modern street argot. "Getting with it," "digging it," "grooving on it" are all slang reflections of this identity. It is this identity that is the basis of craftsmanship in all the technical arts. And it is this identity that modern, dualistically conceived technology lacks. The creator of it feels no particular sense of identity with it. The owner of it feels no particular sense of identity with it. The user of it feels no particular sense of identity with it. Hence, by Phaedrus' definition, it has no Quality.

That wall in Korea that Phaedrus saw was an act of technology. It was beautiful, but not because of any masterful intellectual planning or any scientific supervision of the job, or any added expenditures to "stylize" it. It was beautiful because the people who worked on it had a way of looking at things that made them do it right unself-consciously. They didn't separate themselves from the work in such a way as to do it wrong. There is the center of the whole solution.

The way to solve the conflict between human values and technological needs is not to run away from technology. That's impossible. The way to resolve the conflict is to break down the barriers of dualistic thought that prevent a real understanding of what technology is—not an exploitation of nature, but a fusion of nature and the human spirit into a new kind of creation that transcends both. When this transcendence occurs in such events as the first airplane flight across the ocean or the first footstep on the moon, a kind of public recognition of the transcendent nature of technology occurs. But this transcendence should also occur at the individual level, on a personal basis, in one's own life, in a less dramatic way.

The walls of the canyon here are completely vertical now. In many places room for the road had to be blasted out of it. No alternate routes here. Just whichever way the river goes. It may be just my imagination, but it seems the river's already smaller than it was an hour ago.

Such personal transcendence of conflicts with technology doesn't have to involve motorcycles, of course. It can be at a level as simple as sharpening a kitchen knife or sewing a dress or mending a broken chair. The underlying problems are the same. In each case there's a beautiful way of doing it and an ugly way of doing it, and in arriving at the high-quality, beautiful way of doing it, both an ability to see what "looks good" and an ability to understand the underlying methods to arrive at that "good" are needed. Both classic and romantic understandings of Quality must be combined.

The nature of our culture is such that if you were to look for instruction in how to do any of these jobs, the instruction would always give only one understanding of Quality, the classic. It would tell you how to hold the blade when sharpening the knife, or how to use a sewing machine, or how to mix and apply glue with the presumption that once these underlying methods were applied, "good" would naturally follow. The ability to see directly what "looks good" would be ignored.

The result is rather typical of modern technology, an overall dullness of appearance so depressing that it must be overlaid with a veneer of "style" to make it acceptable. And that, to anyone who is sensitive to romantic Quality, just makes it all the worse. Now it's not just depressingly dull, it's also phony. Put the two together and you get a pretty accurate basic description of modern American technology: stylized cars and stylized outboard motors and stylized typewriters and stylized clothes. Stylized refrigerators filled with stylized food in stylized kitchens in stylized homes. Plastic stylized toys for stylized children, who at Christmas and birthdays are in style with their stylish parents. You have to be awfully stylish yourself not to get sick of it once in a while. It's the style that gets you; technological ugliness syruped over with romantic phoniness in an effort to produce beauty and profit by people who, though stylish, don't know where to start because no one has ever told them there's such a thing as Quality in this world and it's real, not style. Quality isn't something you lay on top of subjects and objects like tinsel on a Christmas tree. Real Quality must be the source of the subjects and objects, the cone from which the tree must start.

To arrive at *this* Quality requires a somewhat different procedure from the "Step 1, Step 2, Step 3" instructions that accompany dualistic technology, and that's what I'll now try to go into.

After many turns in the canyon wall we stop for a break under a scrubby little patch of small trees and rocks. The grass around the trees is burned and brown and scattered with litter from picnickers.

I collapse into some shade, and after a while squint up at the sky, which I haven't really looked at since we entered this canyon. Up there above the canyon walls it's cool and dark blue and far away.

Chris doesn't even go over to see the river, something he'd normally do. Like me, he's tired and content just to sit under the scant shade of these trees.

After a while he says there's an old iron pump, it looks like, between us and the river. He points to it and I see what he means. He goes over and I can see him pump water onto his hand and then splash it onto his face. I go over and pump for him so he can use both hands. Then I do the same. The water feels cold on my hands and face. When done we walk to the cycle again and climb on and pull back on to the canyon road.

Now that solution. Throughout this Chautauqua so far this whole problem of technological ugliness has been looked at in a negative way. It's been said that romantic attitudes toward Quality such as the Sutherlands have are, by themselves, hopeless. You can't live on just groovy emotions alone. You have to work with the underlying form of the universe too, the laws of nature which, when understood, can make work easier, sickness rarer and famine almost absent. On the other hand, technology based on pure dualistic reason has also been condemned because it obtains these material advantages by turning the world into a stylized garbage dump. Now's the time to stop condemning things and come up with some answers.

The answer is to Phaedrus' contention that classic

understanding should not be *overlaid* with romantic prettiness; classic and romantic understanding should be united at a basic level. In the past our common universe of reason has been in the process of escaping, rejecting the romantic, irrational world of prehistoric man. It's been necessary since before the time of Socrates to reject the passions, the emotions, in order to free the rational mind for an understanding of nature's order which was as yet unknown. Now it's time to further an understanding of nature's order by reassimilating those passions which were originally fled from. The passions, the emotions, the affective domain of man's consciousness, are a part of nature's order too. The central part.

At present we're snowed under with an irrational expansion of blind data-gathering in the sciences because there's no rational format for any understanding of scientific creativity. At present we are also snowed under with a lot of stylishness in the arts—thin art—because there's very little assimilation or extension into underlying form. We have artists with no scientific knowledge and scientists with no artistic knowledge and both with no spiritual sense of gravity at all, and the results is not just bad, it is ghastly. The time for real reunification of art and technology is really long overdue.

At the DeWeeses I started to talk about peace of mind in connection with technical work but got laughed off the scene because I brought it up out of the context in which it had originally appeared to me. Now I think it *is* in context to return to peace of mind and see what I was talking about.

Peace of mind isn't at all superficial to technical

work. It's the whole thing. That which produces it is good work and that which destroys it is bad work. The specs, the measuring instruments, the quality control, the final check-out, these are all *means* toward the end of satisfying the peace of mind of those responsible for the work. What really counts in the end is their peace of mind, nothing else. The reason for this is that peace of mind is a prerequisite for a perception of that Quality which is beyond romantic Quality and classic Quality and which unites the two, and which must accompany the work as it proceeds. The way to see what looks good and understand the reasons it looks good, and *to be at one with this goodness* as the work proceeds, is to cultivate an inner quietness, a peace of mind so that goodness can shine through.

I say *inner* peace of mind. It has no direct relationship to external circumstances. It can occur to a monk in meditation, to a soldier in heavy combat or to a machinist taking off that last ten-thousandth of an inch. It involves unself-consciousness, which produces a complete identification with one's circumstances, and there are levels and levels of this identification and levels and levels of quietness quite as profound and difficult of attainment as the more familiar levels of activity. The mountains of achievement are Quality discovered in one direction only, and are relatively meaningless and often unobtainable unless taken together with the ocean trenches of self-awareness—so different from self-consciousness—which result from inner peace of mind.

This inner peace of mind occurs on three levels of understanding. Physical quietness seems the easiest to achieve, although there are levels and levels of this too, as attested by the ability of Hindu mystics to live buried alive for many days. Mental quietness, in which one has no wandering thoughts at all, seems more difficult, but can be achieved. But value quietness, in which one has no wandering desires at all but simply performs the acts of his life without desire, that seems the hardest.

I've sometimes thought this inner peace of mind, this quietness is similar to if not identical with the sort of calm you sometimes get when going fishing, which accounts for much of the popularity of this sport. Just to sit with the line in the water, not moving, not really thinking about anything, not really caring about anything either, seems to draw out the inner tensions and frustrations that have prevented you from solving problems you couldn't solve before and introduced ugliness and clumsiness into your actions and thoughts.

You don't have to go fishing, of course, to fix your motorcycle. A cup of coffee, a walk around the block, sometimes just putting off the job for five minutes of silence is enough. When you do you can almost feel yourself grow toward that inner peace of mind that reveals it all. That which turns its back on this inner calm and the Quality it reveals is bad maintenance. That which turns toward it is good. The forms of turning away and toward are infinite but the goal is always the same.

I think that when this concept of peace of mind is introduced and made central to the act of technical work, a fusion of classic and romantic quality can take place at a basic level within a practical working context. I've said you can actually *see* this fusion in skilled mechanics and machinists of a certain sort, and you can see it in the work they do. To say that they are not artists is to misunderstand the nature of art. They have patience, care and attentiveness to what they're doing, but more than this—there's a kind of inner peace of mind that isn't contrived but results from a kind of harmony with the work in which there's no leader and no follower. The material and the craftsman's thoughts change together in a progression of smooth, even changes until his mind is at rest at the exact instant the material is right.

We've all had moments of that sort when we're doing something we really want to do. It's just that somehow we've gotten into an unfortunate separation of those moments from work. The mechanic I'm talking about doesn't make this separation. One says of him that he is "interested" in what he's doing, that he's "involved" in his work. What produces this involvement is, at the cutting edge of consciousness, an absence of any sense of separateness of subject and object. "Being with it," "being a natural," "taking hold"—there are a lot of idiomatic expressions for what I mean by this absence of subject-object duality, because what I mean is so well understood as folklore, common sense, the everyday understanding of the shop. But in scientific parlance the words for this absence of subject-object duality are scarce because scientific minds have shut themselves off from consciousness of this kind of understanding in the assumption of the formal dualistic scientific outlook.

Zen Buddhists talk about "just sitting," a meditative practice in which the idea of a duality of self and object does not dominate one's consciousness. What I'm talking about here in motorcycle maintenance is "just fixing," in which the idea of a duality of self and object doesn't dominate one's consciousness. When one isn't dominated by feelings of separateness from what he's working on, then one can be said to "care" about what he's doing. That is what caring really is, a feeling of identification with what one's doing. When one has this feeling then he also sees the inverse side of caring, Quality itself.

So the thing to do when working on a motorcycle, as in any other task, is to cultivate the peace of mind which does not separate one's self from one's surroundings. When that is done successfully then everything else follows naturally. Peace of mind produces right values, right values produce right thoughts. Right thoughts produce right actions and right actions produce work which will be a material reflection for others to see of the serenity at the center of it all. That was what it was about that wall in Korea. It was a material reflection of a spiritual reality.

I think that if we are going to reform the world, and make it a better place to live in, the way to do it is not with talk about relationships of a political nature, which are inevitably dualistic, full of subjects and objects and their relationship to one another; or with programs full of things for other people to do. I think that kind of approach starts it at the end and presumes the end is the beginning. Programs of a political nature are important *end products* of social quality that can be effective only if the underlying structure of social values is right. The social values are right only if the individual values are right. The place to improve the world is first in one's own heart and head and hands, and then work outward from there. Other people can talk about how to expand the destiny of mankind. I just want to talk about how to fix a motorcycle. I think that what I have to say has more lasting value.

A town called Riggins comes up and we see a lot of motels, and afterward the road branches away from the canyon and follows a smaller stream. It seems to head upward into forest.

It does, and soon the road becomes shaded by tall, cool pines. Resort signs appear. We wind higher and higher into unexpectedly pleasant, cool, green meadows surrounded by pine forests. At a town called New Meadows we fill up again and buy two cans of oil, still surprised at the change.

But as we leave New Meadows I note the long slant of the sun and a late afternoon depression begins to set in. At another time of day these mountain meadows would refresh me more, but we've gone too long. We pass Tamarack and the road drops down again from green meadows into dry sandy country.

I guess that's all I want to say for the Chautauqua today. It's been a long session and perhaps the most important one. Tomorrow I want to talk about things that seem to turn one toward Quality and turn one away from Quality, some of the traps and problems that come up.

Strange feelings from the orange sunlight on this sandy dry country so far from home. I wonder if Chris feels it too. Just a sort of unexplained sadness that comes each afternoon when the new day is gone forever and there's nothing ahead but increasing darkness.

The orange turns to dull bronze light and continues to show what it has shown all day long, but now it seems to show it without enthusiasm. Across those dry hills, within those little houses in the distance are people who've been there all day long, going about the business of the day, who now find nothing unusual or different in this strange darkening landscape, as we do. If we were to come upon them early in the day they might be curious about us and what we're here for. But now in the evening they'd just resent our presence. The workday is over. It's time for supper and family and relaxation and turning inward at home. We ride unnoticed down this empty highway through this strange country I've never seen before, and now a heavy feeling of isolation and loneliness becomes dominant and my spirits wane with the sun.

We stop at an abandoned school yard and under a huge cottonwood tree I change the oil in the cycle. Chris is irritable and wonders why we stop for so long, not knowing perhaps that it's just the time of day that makes him irritable; but I give him the map to study while I change the oil, and when the oil is changed we look at the map together and decide to have supper at the next good restaurant we find and camp at the first good camping place. That cheers him up.

At a town called Cambridge we have supper and when we are finished, it's dark out. We follow the headlight beam down a secondary road toward Oregon to a little sign saying "BROWNLEE CAMPGROUND," which appears to be in a draw of the mountains. In the dark it's hard to tell what sort of country we're in. We follow a dirt road under trees and past underbrush to some camper's pull-ins. No one else seems to be here.

When I shut the motor off and we unpack I can hear a small stream nearby. Except for that and the chirping of some little bird there's no sound.

"I like it here," Chris says.

"It's very quiet," I say.

"Where will we be going tomorrow?"

"Into Oregon." I give him the flashlight and have him shine it where I'm unpacking.

"Have I been there before?"

"Maybe, I'm not sure."

I spread out the sleeping bags, and put his on top of the picnic table. The novelty of this appeals to him. This night there'll be no trouble sleeping. Soo I hear deep breathing that tells me he's already asleep.

I wish I knew what to say to him. Or what to ask. He seems so close at times, and yet the closeness has nothing to do with what is asked or said. Then at other times he seems very far away and sort of watching me from some vantage point I don't see. And then sometimes he's just childish and there's no relation at all.

Sometimes, when thinking about this, I thought that the idea that one person's mind is accessible to another's is just a conversational illusion, just a figure of speech, an assumption that makes some kind of exchange between basically alien creatures seem plausible, and that really the relationship of one person to another is ultimately unknowable. The effort of fathoming what is in another's mind creates a distortion of what is seen. I'm trying, I suppose, for some situation in which whatever it is emerges undistorted. The way he asks all those questions, I don't know.

26

A sensation of cold wakes me up. I see out the top of the sleeping bag that the sky is dark grey. I pull my head down and close my eyes again.

Later I see the grey of the sky is lighter, and it's still cold. I can see the vapor of my breath. An alarmed thought that the grey is from rain clouds overhead wakes me up, but after looking carefully I see that this is just grey dawn. It seems too cold and early to start riding yet, so I don't get out of the bag. But sleep is gone.

Through the spokes of the motorcycle wheel I see Chris's sleeping bag on the picnic table, twisted all around him. He isn't stirring.

The cycle looms silently over me, ready to start, as if it has waited all night like some silent guardian.

Silver-grey and chrome and black—and dusty. Dirt from Idaho and Montana and the Dakotas and Minnesota. From the ground up it looks very impressive. No frills. Everything with a purpose.

I don't think I'll ever sell it. No reason to, really. They're not like cars, with a body that rusts out in a few years. Keep them tuned and overhauled and they'll last as long as you do. Probably longer. Quality. It's carried us so far without trouble.

The sunlight just touches the top of the bluff high above the draw we're in. A wisp of fog has appeared above the creek. That means it'll warm up.

I get out of the sleeping bag, put shoes on, pack everything I can without waking Chris, and then go over to the picnic table and give him a shake to wake him up.

He doesn't respond. I look around and see that there are no jobs left to do but wake him up, and hesitate, but feeling manic and jumpy from the brisk morning air holler, "WAKE!" and he sits up suddenly, eyes wide open.

I do my best to follow this with the opening Quatrain of *The Rubáiyat of Omar Khayyám.* It looks like some desert cliff in Persia above us. But Chris doesn't know what the hell I'm talking about. He looks up at the top of the bluff and then just sits there squinting at me. You have to be in a certain mood to accept bad recitations of poetry. Particularly that one.

Soon we're on the road again, which twists and turns. We stem down into an enormous canyon with high white bluffs on either side. The wind freezes. The road comes into some sunlight which seems to warm me right through the jacket and sweater, but soon we ride into the shade of the canyon wall again where again the wind freezes. This dry desert air doesn't hold heat. My lips, with the wind blowing into them, feel dry and cracked.

Farther on we cross a dam and leave the canyon into some high semidesert country. This is Oregon now. The road winds through a landscape that reminds me of northern Rajasthan, in India, where it's not quite desert, much piñon, junipers and grass, but not agricultural either, except where a draw or valley provides a little extra water.

Those crazy *Rubáiyat* Quatrains keep rumbling through my head.

> *. . . something, something along some Strip of*
> *Herbage strown,*
> *That just divides the desert from the sown,*
> *Where name of Slave and Sultan scarce is*
> *known,*
> *And pity Sultan Mahmud on his Throne. . . .*

That conjures up a glimpse of the ruins of an ancient Mogul palace near the desert where out of the corner of his eye he saw a wild rosebush. . . .

. . . And this first summer Month that brings the Rose . . . How did that go? I don't know. I don't even *like* the poem. I've noticed since this trip has started and particularly since Bozeman that these fragments seem less and less a part of *his* memory and more and more a part of mine. I'm not sure what that means . . . I think I just don't know.

I think there's a name for this kind of semidesert, but I can't think of what it is. No one can be seen anywhere on the road but us.

Chris hollers that he has diarrhea again. We ride until I see a stream below and pull off the road and stop. His face is full of embarrassment again but I tell him we're in no hurry and get out a change of underwear and roll of toilet paper and bar of soap and tell him to wash his hands thoroughly and carefully after he's done.

I sit on an Omar Khayyám rock contemplating the semidesert and feel not bad.

. . . *And this first Summer month that brings the rose* . . . oh . . . now it comes back . . .

Each Morn a thousand Roses brings, you say,
Yes, but where leaves the Rose of Yesterday?
And this first Summer month that brings the
  Rose,
Shall take Jamshyd and Kaikobad away.

. . . And so on and so forth . . .

Let's get off Omar and onto the Chautauqua. Omar's solution is just to sit around and guzzle the wine and feel so bad that time is passing and the Chautauqua looks good to me by comparison. Particularly today's Chautauqua, which is about gumption.

I see Chris coming back up the hill now. His expression looks happy.

I like the word "gumption" because it's so homely and so forlorn and so out of style it looks as if it needs a friend and isn't likely to reject anyone who comes along. It's an old Scottish word, once used a lot by pi-

placeable" assembly. These drain off gumption, destroy enthusiasm and leave you so discouraged you want to forget the whole business. I call these things "gumption traps."

There are hundreds of different kinds of gumption traps, maybe thousands, maybe millions. I have no way of knowing how many I don't know. I know it *seems* as though I've stumbled onto every kind of gumption trap imaginable. What keeps me from thinking I've hit them all is that with every job I discover more. Motorcycle maintenance gets frustrating. Angering. Infuriating. That's what makes it interesting.

The map before me says the town of Baker is soon ahead. I see we're in better agricultural land now. More rain here.

What I have in mind now is a catalog of "Gumption Traps I Have Known." I want to start a whole new academic field, gumptionology, in which these traps are sorted, classified, structured into hierarchies and interrelated for the edification of future generations and the benefit of all mankind.

Gumptionology 101—An examination of affective, cognitive and psychomotor blocks in the perception of Quality relationships—3 cr, VII, MWF. I'd like to see that in a college catalog somewhere.

In traditional maintenance gumption is considered something you're born with or have acquired as a result of good upbringing. It's a fixed commodity. From the lack of information about how one acquires this gumption one might assume that a person without any gumption is a hopeless case.

In nondualistic maintenance gumption isn't a fixed commodity. It's variable, a reservoir of good spirits that can be added to or subtracted from. Since it's a result of the perception of Quality, a gumption trap, consequently, can be defined as anything that causes one to lose sight of Quality, and thus lose one's enthusiasm for what one is doing. As one might guess from a definition as broad as this, the field is enormous and only a beginning sketch can be attempted here.

As far as I can see there are two main types of gumption traps. The first type is those in which you're thrown off the Quality track by conditions that arise from external circumstances, and I call these "setbacks." The second type is traps in which you're thrown off the Quality track by conditions that are primarily within yourself. These I don't have any generic name for—"hang-ups," I suppose. I'll take up the externally caused setbacks first.

The first time you do any major job it seems as though the out-of-sequence-reassembly setback is your biggest worry. This occurs usually at a time when you think you're almost done. After days of work you finally have it all together except for: What's this? *A connecting-rod bearing liner?!* How could you have left *that* out? Oh Jesus, everything's got to come *apart* again! You can almost hear the gumption escaping. *Pssssssssssssss.*

There's nothing you can do but go back and take it all apart again . . . after a rest period of up to a month that allows you to get used to the idea.

There are two techniques I use to prevent the out-of-sequence-reassembly setback. I use them mainly when

I'm getting into a complex assembly I don't know anything about.

It should be inserted here parenthetically that there's a school of mechanical thought which says I shouldn't *be* getting into a complex assembly I don't know anything about. I should have training or leave the job to a specialist. That's a self-serving school of mechanical eliteness I'd like to see wiped out. That was a "specialist" who broke the fins on this machine. I've edited manuals written to train specialists for IBM, and what they know when they're done isn't that great. You're at a disadvantage the first time around and it may cost you a little more because of parts you accidently damage, and it will almost undoubtedly take a lot more time, but the next time around you're way ahead of the specialist. You, with gumption, have learned the assembly the hard way and you've a whole set of good feelings about it that he's unlikely to have.

Anyway, the first technique for preventing the out-of-sequence-reassembly gumption trap is a notebook in which I write down the order of disassembly and note anything unusual that might give trouble in reassembly later on. This notebook gets plenty grease-smeared and ugly. But a number of times one or two words in it that didn't seem important when written down have prevented damage and saved hours of work. The notes should pay special attention to left-hand and right-hand and up-and-down orientations of parts, and color coding and positions of wires. If incidental parts look worn or damaged or loose this is the time to note it so that you can make all your parts purchases at the same time.

The second technique for preventing the out-of-sequence-reassembly gumption trap is newspapers opened out on the floor of the garage on which all the parts are laid left-to-right and top-to-bottom in the order in which you read a page. That way when you put it back together in reverse order the little screws and washers and pins that can be easily overlooked are brought to your attention as you need them.

Even with all these precautions, however, out-of-sequence reassemblies sometimes occur and when they do you've got to watch the gumption. Watch out for gumption desperation, in which you hurry up wildly in an effort to restore gumption by making up for lost time. That just creates more mistakes. When you first see that you have to go back and take it apart all over again it's definitely time for that long break.

It's important to distinguish from these the reassemblies that were out of sequence because you lacked certain information. Frequently the whole reassembly process becomes a cut-and-try technique in which you have to take it apart to make a change and then put it together again to see if the change works. If it doesn't work, that isn't a setback because the information gained is a real progress.

But if you've made just a plain old dumb mistake in reassembly, some gumption can still be salvaged by the knowledge that the second disassembly and reassembly is likely to go much faster than the first one. You've unconsciously memorized all sorts of things you won't have to relearn.

From Baker the cycle has taken us up through forests. The forest road takes us through a pass and down through more forests on the other side.

As we move again down the side of the mountain we see the trees thin out even more until we are in desert again.

The intermittent failure setback is next. In this the thing that is wrong becomes right all of a sudden just as you start to fix it. Electrical short circuits are often in this class. The short occurs only when the machine's bouncing around. As soon as you stop everything's okay. It's almost impossible to fix it then. All you can do is try to get it to go wrong again and if it won't, forget it.

Intermittents become gumption traps when they fool you into thinking you've really got the machine fixed. It's always a good idea on any job to wait a few hundred miles before coming to that conclusion. They're discouraging when they crop up again and again, but when they do you're no worse off than someone who goes to a commercial mechanic. In fact you're better off. They're much more of a gumption trap for the owner who has to drive his machine to the shop again and again and never get satisfaction. On your own machine you can study them over a long period of time, something a commercial machine can't do, and you can just carry around the tools you think you'll need until the intermittent happens again, and then, when it happens, stop and work on it.

When intermittents recur, try to correlate them with other things the cycle is doing. Do the misfires, for example, occur only on bumps, only on turns, only on acceleration? Only on hot days? These correlations are clues for cause-and-effect hypotheses. In some intermittents you have to resign yourself to a long fishing expedition, but no matter how tedious that gets it's never as tedious as taking the machine to a commercial mechanic five times. I'm tempted to go into long detail about "Intermittents I Have Known" with a blow-by-blow description of how these were solved. But this gets like those fishing stories, of interest mainly to the fisherman, who doesn't quite catch on to why everybody yawns. *He* enjoyed it.

Next to misassemblies and intermittents I think the most common external gumption trap is the parts setback. Here a person who does his own work can get depressed in a number of ways. Parts are something you never plan on buying when you originally get the machine. Dealers like to keep their inventories small. Wholesalers are slow and always understaffed in the spring when everybody buys motorcycle parts.

The pricing on parts is the second part of this gumption trap. It's a well-known industrial policy to price the original equipment competitively, because the customer can always go somewhere else, but on parts to overprice and clean up. The price of the part is not only jacked up way beyond its new price; you get a special price because you're not a commercial mechanic. This is a sly arrangement that allows the commercial mechanic to get rich by putting in parts that aren't needed.

One more hurdle yet. The part may not fit. Parts lists always contain mistakes. Make and model changes are confusing. Out-of-tolerance parts runs sometimes get through quality control because there's no operating checkout at the factory. Some of the parts you buy are made by specialty houses who don't have access to the engineering data needed to make them right. Sometimes *they* get confused about make and model changes. Sometimes the parts man you're dealing with jots down the wrong number. Sometimes you don't give him the right identification. But it's always a major gumption trap to get all the way home and discover that a new part won't work.

The parts traps may be overcome by a combination of a number of techniques. First, if there's more than one supplier in town by all means choose the one with the most cooperative parts man. Get to know him on a first-name basis. Often he will have been a mechanic once himself and can provide a lot of information you need.

Keep an eye out for price cutters and give them a try. Some of them have good deals. Auto stores and mail-order houses frequently stock the commoner cycle parts at prices way below those of the cycle dealers. You can buy roller chain from chain manufacturers, for example, at way below the inflated cycle-shop prices.

Always take the old part with you to prevent getting a wrong part. Take along some machinist's calipers for comparing dimensions.

Finally, if you're as exasperated as I am by the parts problem and have some money to invest, you can take up the really fascinating hobby of machining your own parts. I have a little 6-by-18-inch lathe with a milling attachment and a full complement of welding equipment: arc, heli-arc, gas and mini-gas for this kind of work. With the welding equipment you can build up worn surfaces with better than original metal and then machine it back to tolerance with carbide tools. You can't really believe how versatile that lathe-plus-milling-plus-welding arrangement is until you've used it. If you can't do the job directly you can always make something that *will* do it. The work of machining a part is very slow, and some parts, such as ball bearings, you're never going to machine, but you'd be amazed at how you can modify parts designs so that you can make them with your equipment, and the work isn't nearly as slow or frustrating as a wait for some smirking parts man to send away to the factory. And the work is gumption *building,* not gumption destroying. To run a cycle with parts in it you've made yourself gives you a special feeling you can't possibly get from strictly store-bought parts.

We've come into the sage and sand of the desert and the engine's started to sputter. I switch to the reserve gas tank and study the map. We fill up at a town called Unity and down the hot black road, through the sagebrush we go.

Well, those were the commonest setbacks I can think of: out-of-sequence reassembly, intermittent failure and parts problems. But although setbacks are the commonest gumption traps they're only the external cause of gumption loss. Time now to consider some of the *internal* gumption traps that operate at the same time.

As the course description of gumptionology indicated, this internal part of the field can be broken down into three main types of internal gumption traps: those that block affective understanding, called "value traps"; those that block cognitive understanding, called "truth traps"; and those that block psychomotor behavior, called "muscle traps." The value traps are by far the largest and the most dangerous group.

Of the value traps, the most widespread and pernicious is value rigidity. This is an inability to revalue what one sees because of commitment to previous values. In motorcycle maintenance, you *must* rediscover what you do as you go. Rigid values make this impossible.

The typical situation is that the motorcycle doesn't work. The facts are there but you don't see them. You're looking right at them, but they don't yet have enough *value*. This is what Phaedrus was talking about. Quality, value, *creates* the subjects and objects of the world. The facts do not exist until value has created them. If your values are rigid you can't really learn new facts.

This often shows up in premature diagnosis, when you're sure you know what the trouble is, and then when it isn't, you're stuck. Then you've got to find some new clues, but before you can find them you've got to clear your head of old opinions. If you're plagued with value rigidity you can fail to see the real answer even when it's staring you right in the face because you can't see the new answer's importance.

The birth of a new fact is always a wonderful thing to experience. It's dualistically called a "discovery" because of the presumption that it has an existence independent of anyone's awareness of it. When it comes along, it always has, at first, a low value. Then, depending on the value-looseness of the observer and the potential quality of the fact, its value increases, either slowly or rapidly, or the value wanes and the fact disappears.

The overwhelming majority of facts, the sights and sounds that are around us every second and the relationships among them and everything in our memory—these have no Quality, in fact have a negative quality. If they were all present at once our consciousness would be so jammed with meaningless data we couldn't think or act. So we preselect on the basis of Quality, or, to put it Phaedrus' way, the track of Quality preselects what data we're going to be conscious of, and it makes this selection in such a way as to best harmonize what we are with what we are becoming.

What you have to do, if you get caught in this gumption trap of value rigidity, is slow down—you're going to have to slow down anyway whether you want to or not—but slow down deliberately and go over ground that you've been over before to see if the things you thought were important were really important and to . . . well . . . just *stare* at the machine. There's nothing wrong with that. Just live with it for a while. Watch it the way you watch a line when fishing and before long, as sure as you live, you'll get a little nibble, a little fact asking in a timid, humble way if you're interested in it. That's the way the world keeps on happening. Be interested in it.

At first try to understand this new fact not so much in terms of your big problem as for its own sake. That problem may not be as big as you think it is. And that fact may not be as small as you think it is. It may not be the fact you want but at least you should be very sure of that before you send the fact away. Often before you send it away you will discover it has friends who are right next to it and are watching to see what your response is. Among the friends may be the exact fact you are looking for.

After a while you may find that the nibbles you get

are more interesting than your original purpose of fixing the machine. When that happens you've reached a kind of point of arrival. Then you're no longer strictly a motorcycle mechanic, you're also a motorcycle scientist, and you've completely conquered the gumption trap of value rigidity.

The road has come up into the pines again, but I see by the map that it won't be for long. There are some resort billboards along the road and some kids beneath them, almost as if part of the advertisement, gathering pinecones. They wave and in doing so the littlest boy drops all his cones.

I keep wanting to go back to that analogy of fishing for facts. I can just see somebody asking with great frustration, "Yes, but *which* facts do you fish for? There's got to be more to it than *that*."

But the answer is that if you know which facts you're fishing for you're no longer fishing. You've caught them. I'm trying to think of a specific example . . .

All kinds of examples from cycle maintenance could be given, but the most striking example of value rigidity I can think of is the old South Indian Monkey Trap, which depends on value rigidity for its effectiveness. The trap consists of a hollowed-out coconut chained to a stake. The coconut has some rice inside which can be grabbed through a small hole. The hole is big enough so that the monkey's hand can go in, but too small for his fist with rice in it to come out. The monkey reaches in and is suddenly trapped—by nothing more than his own value rigidity. He can't revalue the rice. He cannot see that freedom without rice is more valuable than capture with it. The villagers are coming to get him and take him away. They're coming closer . . . closer! . . . now! What general advice—not specific advice—but what *general* advice would you give the poor monkey in circumstances like this?

Well, I think you might say exactly what I've been saying about value rigidity, with perhaps a little extra urgency. There is a fact this monkey should know: if he opens his hand he's free. But how is he going to discover this fact? By removing the value rigidity that rates rice above freedom. How is he going to do that? Well, he should somehow try to slow down deliberately and go over ground that he has been over before and see if things he thought were important really *were* important and, well, stop yanking and just stare at the coconut for a while. Before long he should get a nibble from a little fact wondering if he is interested in it. He should try to understand this fact not so much in terms of his big problem as for its own sake. That problem may not be as big as he thinks it is. That fact may not be as small as he thinks it is either. That's about all the general information you can give him.

At Prairie City we're out of the mountain forests again and into a dryland town with a wide main street that looks right down through the center of the town and onto the prairie beyond it. We try one restaurant, but it's closed. We go across the broad street and try another. The door's open, we sit down and order malted milks. While waiting I get the outline of the letter Chris was preparing for his mother and give it to him.

To my surprise he works on it without many questions. I sit back in the booth and don't disturb him.

I keep feeling that the facts I'm fishing for concerning Chris are right in front of me too, but that some value rigidity of my own blocks me from seeing it. At times we seem to move in parallel rather than in combination, then at odd moments collide.

His troubles at home always begin when he is imitating me, trying to command others the way I command him, particularly his younger brother. Naturally the others aren't having any of his commands, and he can't see their right not to, and that's when all hell breaks loose.

He can't seem to care whether he's popular with anyone else. He just wants to be popular with me. Not healthy at all, everything considered. It's about time for him to begin the long process of breaking away. That break should be as easy as possible, but it should be made. It's time to set him on his own feet. The sooner the better.

And now, having thought all that, I don't believe it anymore. I don't know what the trouble is. That dream that keeps recurring haunts me because I can't escape its meaning: I'm forever on the other side of a glass door from him which I don't open. He wants me to open it and before I always turned away. But now there's a new figure who prevents me. Strange.

After a while Chris says he's tired of writing. We get up, I pay at the counter and we leave.

On the road now and talking about traps again.

The next one is important. It's the internal gumption

trap of ego. Ego isn't entirely separate from value rigidity but one of the many causes of it.

If you have a high evaluation of yourself then your ability to recognize new facts is weakened. Your ego isolates you from the Quality reality. When the facts show that you've just goofed, you're not as likely to admit it. When false information makes you look good, you're likely to believe it. On any mechanical repair job ego comes in for rough treatment. You're always being fooled, you're always making mistakes, and a mechanic who has a big ego to defend is at a terrific disadvantage. If you know enough mechanics to think of them as a group, and your observations coincide with mine, I think you'll agree that mechanics tend to be rather modest and quiet. There are exceptions, but generally if they're not quiet and modest at first, the work seems to make them that way. And skeptical. Attentive, but skeptical. But not egoistic. There's no way to bullshit your way into looking good on a mechanical repair job, except with someone who doesn't know what you're doing.

. . . I was going to say that the machine doesn't respond to your personality, but it *does* respond to your personality. It's just that the personality that it responds to is your *real* personality, the one that genuinely feels and reasons and acts, rather than any false, blown-up personality images your ego may conjure up. These false images are deflated so rapidly and completely you're bound to be very discouraged very soon if you've derived your gumption from ego rather than Quality.

If modesty doesn't come easily or naturally to you, one way out of this trap is to fake the attitude of modesty anyway. If you just deliberately assume you're not much good, then your gumption gets a boost when the facts prove this assumption is correct. This way you can keep going until the time comes when the facts prove this assumption is *in*correct.

Anxiety, the next gumption trap, is sort of the opposite of ego. You're so sure you'll do everything wrong you're afraid to do anything at all. Often this, rather than "laziness," is the real reason you find it hard to get started. This gumption trap of anxiety, which results from overmotivation, can lead to all kinds of errors of excessive fussiness. You fix things that don't need fixing, and chase after imaginary ailments. You jump to wild conclusions and build all kinds of errors into the machine because of your own nervousness. These errors, when made, tend to confirm your original underestimation of yourself. This leads to more errors, which lead to more underestimation, in a self-stoking cycle.

The best way to break this cycle, I think, is to work out your anxieties on paper. Read every book and magazine you can on the subject. Your anxiety makes this easy and the more you read the more you calm down. You should remember that it's peace of mind you're after and not just a fixed machine.

When beginning a repair job you can list everything you're going to do on little slips of paper which you then organize into proper sequence. You discover that you organize and then reorganize the sequence again and again as more and more ideas come to you. The time spent this way usually more than pays for itself in time saved on the machine and prevents you from doing fidgety things that create problems later on.

You can reduce your anxiety somewhat by facing the fact that there isn't a mechanic alive who doesn't louse up a job once in a while. The main difference between you and the commercial mechanics is that when they do it you don't hear about it—just pay for it, in additional costs prorated through all your bills. When you make the mistakes yourself, you at least get the benefit of some education.

Boredom is the next gumption trap that comes to mind. This is the opposite of anxiety and commonly goes with ego problems. Boredom means you're off the Quality track, you're not seeing things freshly, you've lost your "beginner's mind" and your motorcycle is in great danger. Boredom means your gumption supply is low and must be replenished before anything else is done.

When you're bored, *stop*! Go to a show. Turn on the TV. Call it a day. Do anything but work on that machine. If you don't stop, the next thing that happens is the Big Mistake, and then all the boredom plus the Big Mistake combine together in one Sunday punch to knock all the gumption out of you and you are really stopped.

My favorite cure for boredom is sleep. It's very easy to get to sleep when bored and very hard to get bored after a long rest. My next favorite is coffee. I usually keep a pot plugged in while working on the machine. If these don't work it may mean deeper Quality problems are bothering you and distracting you from what's before you. The boredom is a signal that you should turn your attention to these problems—that's what you're doing anyway—and control them before continuing on the motorcycle.

For me the most boring task is cleaning the machine.

It seems like such a waste of time. It just gets dirty again the first time you ride it. John always kept his BMW spic and span. It really did look nice, while mine's always a little ratty, it seems. That's the classical mind at work, runs fine inside but looks dingy on the surface.

One solution to boredom on certain kinds of jobs such as greasing and oil changing and tuning is to turn them into a kind of ritual. There's an esthetic to doing things that are unfamiliar and another esthetic to doing things that are familiar. I have heard that there are two kinds of welders: production welders, who don't like tricky setups and enjoy doing the same thing over and over again; and maintenance welders, who hate it when they have to do the same job twice. The advice was that if you hire a welder make sure which kind he is, because they're not interchangeable. I'm in that latter class, and that's probably why I enjoy troubleshooting more than most and dislike cleaning more than most. But I can do both when I have to and so can anyone else. When cleaning I do it the way people go to church—not so much to discover anything new, although I'm alert for new things, but mainly to reacquaint myself with the familiar. It's nice sometimes to go over familiar paths.

Zen has something to say about boredom. Its main practice of "just sitting" has got to be the world's most boring activity—unless it's that Hindu practice of being buried alive. You don't do anything much; not move, not think, not care. What could be more boring? Yet in the center of all this boredom is the very thing Zen Buddhism seeks to teach. What is it? What is it at the very center of boredom that you're not seeing?

Impatience is close to boredom but always results from one cause: an underestimation of the amount of time the job will take. You never really know what will come up and very few jobs get done as quickly as planned. Impatience is the first reaction against a setback and can soon turn to anger if you're not careful.

Impatience is best handled by allowing an indefinite time for the job, particularly new jobs that require unfamiliar techniques; by doubling the allotted time when circumstances force time planning; and by scaling down the scope of what you want to do. Overall goals must be scaled down in importance and immediate goals must be scaled up. This requires value flexibility, and the value shift is usually accompanied by some loss of gumption, but it's a sacrifice that must be made. It's nothing like the loss of gumption that will occur if a Big Mistake caused by impatience occurs.

My favorite scaling-down exercise is cleaning up nuts and bolts and studs and tapped holes. I've got a phobia about crossed or jimmied or rust-jammed or dirt-jammed threads that cause nuts to turn slow or hard; and when I find one, I take its dimensions with a thread gauge and calipers, get out the taps and dies, recut the threads on it, then examine it and oil it and I have a whole new perspective on patience. Another one is cleaning up tools that have been used and not put away and are cluttering up the place. This is a good one because one of the first warning signs of impatience is frustration at not being able to lay your hand on the tool you need right away. If you just stop and put tools away neatly you will both find the tool and also scale down your impatience without wasting time or endangering the work.

* * *

We're pulling into Dayville and my rear end feels like it's turned to concrete.

Well, that about does it for value traps. There's a whole lot more of them, of course. I've really only just touched on the subject to show what's there. Almost any mechanic could fill you in for hours on value traps he's discovered that I don't know anything about. You're bound to discover plenty of them for yourself on almost every job. Perhaps the best single thing to learn is to recognize a value trap when you're in it and work on that before you continue on the machine.

Dayville has huge shade trees by the filling station where we wait for the attendant to appear. None does, and we, being stiff and uneager to get back on the cycle, do leg exercises under the shade of the trees. Big trees that almost completely cover the road. Odd, in this desertlike country.

The attendant still doesn't show, but his competitor at the filling station across the narrow intersection is watching this, and soon comes over to fill the tank. "I don't know where John is," he says.

When John appears, he thanks the other attendant and says proudly, "We always help each other out like this."

I ask him if there's a place to rest and he says, "You can use my front lawn." He points across the main road to his house behind some cottonwood trees that must be three to four feet in diameter.

We do this, stretch out on some long green grass, and I see that the grass and trees are irrigated from a ditch by the road that has clear moving water in it.

We must have slept half an hour when we see John is in a rocking chair on the green grass beside us, talking to a fire warden in another chair. I listen. The conversation's pace intrigues me. It isn't intended to go anywhere, just fill the time of day. I haven't heard steady slow-paced conversation like that since the thirties when my grandfather and great-grandfather and uncles and great-uncles used to talk like that: on and on and on with no point or purpose other than to fill time, like the rocking of a chair.

John sees I'm awake and we talk a little. He says the irrigation water comes from the "Chinaman's Ditch." "You never could get a white man to dig a ditch like that," he says. "They dug that ditch eighty years ago when they thought there was gold here. You couldn't get a ditch like that anywhere nowadays." He says that's why the trees are so big.

We talk some about where we're from and where we're going, and when we leave John says he's happy to have met us and hopes we're rested. As we move off under the big trees Chris waves and he smiles and waves back.

The desert road winds through rocky gorges and hills. This is the driest country yet.

I want to talk now about truth traps and muscle traps and then stop this Chautauqua for today.

Truth traps are concerned with data that are apprehended and are within the boxcars of the train. For the most part these data are properly handled by conventional dualistic logic and the scientific method talked about earlier, back just after Miles City. But there's one trap that isn't—the truth trap of yes-no logic.

Yes and no . . . this or that . . . one or zero. On the basis of this elementary two-term discrimination, all human knowledge is built up. The demonstration of this is the computer memory which stores all its knowledge in the form of binary information. It contains ones and zeros, that's all.

Because we're unaccustomed to it, we don't usually see that there's a third possible logical term equal to yes and no which is capable of expanding our understanding in an unrecognized direction. We don't even have a term for it, so I'll have to use the Japanese *mu*.

Mu means "no thing." Like "Quality" it points outside the process of dualistic discrimination. *Mu* simply says, "No class; not one, not zero, not yes, not no." It states that the context of the question is such that a yes or no answer is in error and should not be given. "Unask the question" is what it says.

Mu becomes appropriate when the context of the question becomes too small for the truth of the answer. When the Zen monk Joshu was asked whether a dog had a Buddha nature he said *"Mu,"* meaning that if he answered either way he was answering incorrectly. The Buddha nature cannot be captured by yes-or-no questions.

That *mu* exists in the natural world investigated by science is evident. It's just that, as usual, we're trained not to see it by our heritage. For example, it's stated over and over again that computer circuits exhibit only two states, a voltage for "one" and a voltage for "zero." That's silly!

Any computer-electronics technician knows otherwise. Try to find a voltage representing one or zero when the power is off! The circuits are in a *mu* state.

They aren't at one, they aren't at zero, they're in an indeterminate state that has no meaning in terms of ones or zeros. Readings of the voltmeter will show, in many cases, "floating ground" characteristics, in which the technician isn't reading characteristics of the computer circuits at all but characteristics of the voltmeter itself. What's happened is that the power-off condition is part of a context larger than the context in which the one-zero states are considered universal. The question of one or zero has been "unasked." And there are plenty of other computer conditions besides a power-off condition in which *mu* answers are found because of larger contexts than the one-zero universality.

The dualistic mind tends to think of *mu* occurrences in nature as a kind of contextual cheating, or irrelevance, but *mu* is found throughout all scientific investigation, and nature doesn't cheat, and nature's answers are never irrelevant. It's a great mistake, a kind of dishonesty, to sweep nature's *mu* answers under the carpet. Recognition and valuation of these answers would do a lot to bring logical theory closer to experimental practice. Every laboratory scientist knows that very often his experimental results provide *mu* answers to the yes-no questions the experiments were designed for. In these cases he considers the experiment poorly designed, chides himself for stupidity and at best considers the "wasted" experiment which has provided the *mu* answer to be a kind of wheel-spinning which might help prevent mistakes in the design of future yes-no experiments.

This low evaluation of the experiment which provided the *mu* answer isn't justified. The *mu* answer is an important one. It's told the scientist that the context of his question is too small for nature's answer and that he must enlarge the context of the question. That is a very *important* answer! His understanding of nature is tremendously improved by it, which was the purpose of the experiment in the first place. A very strong case can be made for the statement that science grows by its *mu* answers *more* than by its yes or no answers. Yes or no confirms or denies a hypothesis. *Mu* says the answer is *beyond* the hypothesis. *Mu* is the "phenomenon" that inspires scientific inquiry in the first place! There's nothing mysterious or esoteric about it. It's just that our culture has warped us to make a low value judgment of it.

In motorcycle maintenance the *mu* answer given by the machine to many of the diagnostic questions put to it is a major cause of gumption loss. It *shouldn't* be! When your answer to a test is indeterminate it means one of two things: that your test procedures aren't doing what you think they are or that your understanding of the context of the question needs to be enlarged. Check your tests and restudy the question. Don't throw away those *mu* answers! They're every bit as vital as the yes or no answers. They're *more* vital. They're the ones you *grow* on!

. . . This motorcycle seems to be running a little hot . . . but I suppose it's just the hot dry country we're going through . . . I'll leave the answer to that in a *mu* state . . . until it gets worse or better. . . .

We stop for a long chocolate malted in the town of Mitchell, nestled in some dry hills that we can see out the plate-glass window. Some kids come in on a truck and stop and all pile off and come into the restaurant and sort of dominate it. They're reasonably well-behaved, just noisy and energetic, but you can see the lady who's running it is a little nervous about them.

Dry desert, sandy country again. Into it we go. It's late afternoon now and we've really covered the miles. I'm getting quite sore from sitting all this time on the cycle. Feeling really tired now. So was Chris back at the restaurant. A little despondent too. I think maybe he . . . well . . . let it go. . . .

The *mu* expansion is the only thing I want to say about truth traps at this time. Time to switch to the psychomotor traps. This is the domain of understanding which is most directly related to what happens to the machine.

Here by far the most frustrating gumption trap is inadequate tools. Nothing's quite so demoralizing as a tool hang-up. Buy good tools as you can afford them and you'll never regret it. If you want to save money don't overlook the newspaper want ads. Good tools, as a rule, don't wear out, and good secondhand tools are much better than inferior new ones. Study the tool catalogs. You can learn a lot from them.

Apart from bad tools, bad surroundings are a major gumption trap. Pay attention to adequate lighting. It's amazing the number of mistakes a little light can prevent.

Some physical discomfort is unpreventable, but a lot of it, such as that which occurs in surroundings that are too hot or too cold, can throw your evaluations way off if you aren't careful. If you're too cold, for example, you'll hurry and probably make mistakes. If you're too hot your anger threshold gets much lower. Avoid out-of-position work when possible. A small stool on either side of the cycle will increase your patience greatly and you'll be much less likely to damage the assemblies you're working on.

There's one psychomotor gumption trap, muscular insensitivity, which accounts for some real damage. It results in part from lack of kinesthesia, a failure to realize that although the externals of a cycle are rugged, inside the engine are delicate precision parts which can be easily damaged by muscular insensitivity. There's what's called "mechanic's feel," which is very obvious to those who know what it is, but hard to describe to those who don't; and when you see someone working on a machine who doesn't have it, you tend to suffer with the machine.

The mechanic's feel comes from a deep inner kinesthetic feeling for the elasticity of materials. Some materials, like ceramics, have very little, so that when you thread a porcelain fitting you're very careful not to apply great pressures. Other materials, like steel, have tremendous elasticity, more than rubber, but in a range in which, unless you're working with large mechanical forces, the elasticity isn't apparent.

With nuts and bolts you're in the range of large mechanical forces and you should understand that within these ranges metals are elastic. When you take up a nut there's a point called "finger-tight" where there's contact but no takeup of elasticity. Then there's "snug," in which the easy surface elasticity is taken up. Then there's a range called "tight," in which all the elasticity is taken up. The force required to reach these three points is different for each size of nut and bolt, and different for lubricated bolts and for locknuts. The forces are different for steel and cast iron and brass and aluminum and plastics and ceramics. But a person with mechanic's feel knows when something's tight and stops. A person without it goes right on past and strips the threads or breaks the assembly.

A "mechanic's feel" implies not only an understanding for the elasticity of metal but for its softness. The insides of a motorcycle contain surfaces that are precise in some cases to as little as one ten-thousandth of an inch. If you drop them or get dirt on them or scratch them or bang them with a hammer they'll lose that precision. It's important to understand that the metal *behind* the surfaces can normally take great shock and stress but that the surfaces themselves cannot. When handling precision parts that are stuck or difficult to manipulate, a person with mechanic's feel will avoid damaging the surfaces and work with his tools on the nonprecision surfaces of the same part whenever possible. If he must work on the surfaces themselves, he'll always use softer surfaces to work them with. Brass hammers, plastic hammers, wood hammers, rubber hammers and lead hammers are all available for this work. Use them. Vise jaws can be fitted with plastic and copper and lead faces. Use these too. Handle precision parts gently. You'll never be sorry. If you have a tendency to bang things around, take more time and try to develop a little more respect for the accomplishment that a precision part represents.

Low-angled shadows in the dry country we've been through have left a kind of blue depressed feeling. . . .

*   *   *

Maybe it's just the usual late afternoon letdown, but after all I've said about all these things today I just have a feeling that I've somehow talked *around* the point. Some could ask, "Well, if I get around all those gumption traps, then will I have the thing licked?"

The answer, of course, is no, you still haven't got anything licked. You've got to live right too. It's the way you live that predisposes you to avoid the traps and see the right facts. You want to know how to paint a perfect painting? It's easy. Make yourself perfect and then just paint naturally. That's the way all the experts do it. The making of a painting or the fixing of a motorcycle isn't separate from the rest of your existence. If you're a sloppy thinker the six days of the week you aren't working on your machine, what trap avoidances, what gimmicks, can make you all of a sudden sharp on the seventh? It all goes together.

But if you're a sloppy thinker six days a week and you really *try* to be sharp on the seventh, then maybe the next six days aren't going to be quite as sloppy as the preceding six. What I'm trying to come up with on these gumption traps, I guess, is shortcuts to living right.

The real cycle you're working on is a cycle called yourself. The machine that appears to be "out there" and the person that appears to be "in here" are not two separate things. They grow toward Quality or fall away from Quality together.

We arrive in Prineville Junction with only a few hours of daylight left. We're at the intersection with Highway 97, where we'll turn south, and I fill up the tank at the corner and then am so tired I go around in back and sit on the yellow-painted cement curb with my feet in the gravel and the last rays of the sun flaring through the trees into my eyes. Chris comes and sits down too, and we don't say anything, but this is the worst depression yet. All that talk about gumption traps and I fall right into one myself. Fatigue maybe. We've got to get some sleep.

I watch the cars go by for a while on the highway. Something lonely about them. Not lonely—worse. Nothing. Like the attendant's expression when he filled the tank. Nothing. A nothing curb, by some nothing gravel, at a nothing intersection, going nowhere.

Something about the car drivers too. They look just like the gasoline attendant, staring straight ahead in some private trance of their own. I haven't seen that since . . . since Sylvia noticed it the first day. They all look like they're in a funeral procession.

Once in a while one gives a quick glance and then looks away expressionlessly, as if minding his own business, as if embarrassed that we might have noticed he was looking at us. I see it now because we've been away from it for a long time. The driving is different too. The cars seem to be moving at a steady maximum speed for in-town driving, as though they want to get somewhere, as though what's here right now is just something to get through. The drivers seem to be thinking about where they want to be rather than where they are.

I know what it is! We've arrived at the West Coast! We're all strangers again! Folks, I just forgot the biggest gumption trap of all. The funeral procession! The one everybody's in, this hyped-up, fuck-you, supermodern, ego style of life that thinks it owns this country. We've been out of it for so long I'd forgotten all about it.

We get into the stream of traffic going south and I can feel the hyped-up danger close in. I see in the mirror some bastard is tailgating me and won't pass. I move it up to seventy-five and he still hangs in there. Ninety-five and we pull away from him. I don't like this at all.

At Bend we stop and have supper in a modern restaurant in which people also come and go without looking at each other. The service is excellent but impersonal.

Farther south we find a forest of scrubby trees, subdivided into ridiculous little lots. Some developer's scheme apparently. At one of the lots far off the main highway we spread out our sleeping bags and discover that the pine needles just barely cover what must be many feet of soft spongy dust. I've never seen anything like it. We have to be careful not to kick up the needles or the dust flies up over everything.

We spread out the tarps and put the sleeping bags on them. That seems to work. Chris and I talk for a while about where we are and where we are going. I look at the map in the twilight, and then look at it some more with the flashlight. We've covered 325 miles today. That's a lot. Chris seems as completely tired as I am, and as ready as I am to fall asleep.

PART IV

27

Why don't you come out of the shadows? What do you really look like? You're afraid of something, aren't you? What is it you're afraid of?

Beyond the figure in the shadows is the glass door. Chris is behind it, motioning me to open it. He's older now, but his face still has a pleading expression. "What do I do now?" he wants to know. "What do I do next?" He's waiting for my instructions.

It's time to act.

I study the figure in the shadows. It's not as omnipotent as it once seemed. "Who are you?" I ask.

No answer.

"By what right is that door closed?"

Still no answer. The figure is silent, but it is also cowering. It's afraid! Of *me*.

"There are worse things than hiding in the shadows. Is that it? Is that why you don't speak?"

It seems to be quivering, retreating, as though sensing what I am about to do.

I wait, and then move closer to it. Loathsome, dark, evil thing. Closer, looking not at it but at the glass door, so as not to warm it. I pause again, brace myself and then lunge!

My hands sink into something soft where its neck should be. It writhes, and I tighten the grip, as one holds a serpent. And now holding it tighter and tighter we'll get it into the light. Here it comes! *Now we'll see its face!*

"Dad!"

"Dad!" I hear Chris's voice through the door?

Yes! The first time! "Dad! Dad!"

"Dad! Dad!" Chris tugs on my shirt. "Dad! Wake up! Dad!"

He's crying, sobbing now. "Stop, Dad! Wake up!"

"It's all right, Chris."

"Dad! Wake up!"

"I'm awake." I can just barely make out his face in the dawn light. We're in trees somewhere outside. There's a motorcycle here. I think we're in Oregon somewhere.

"I'm all right, it was just a nightmare."

He continues to cry and I sit quietly with him for a while. "It's all right," I say, but he doesn't stop. He's badly frightened.

So am I.

"What were you dreaming about?"

"I was trying to see someone's face."

"You shouted you were going to kill me."

"No, not you."

"Who?"

"The person in the dream."

"Who was it?"

"I'm not sure."

Chris's crying stops, but he continues to shake from the cold. "Did you see the face?"

"Yes."

"What did it look like?"

"It was my own face, Chris, that's when I shouted. . . . It was just a bad dream." I tell him he's shivering and should get back into the sleeping bag.

He does this. "It's so cold," he says.

"Yes." By the dawn light I can see the vapor from our breaths. Then he crawls under the cover of the sleeping bag and I can see only my own.

I don't sleep.

The dreamer isn't me at all.

It's Phaedrus.

He's waking up.

A mind divided against itself . . . me . . . I'm the evil figure in the shadows. I'm the loathsome one. . . .

I always knew he would come back. . . .

It's a matter now of preparing for it. . . .

The sky under the trees looks so grey and so hopeless.

Poor Chris.

28

The despair grows now.

Like one of those movie dissolves in which you
know you're not in the real world but it seems that way
anyway.

It's a cold, snowless November day. The wind blows
dirt through the cracks of the windows of an old car
with soot on the windows, and Chris, six, sits beside
him, with sweaters on because the heater doesn't work,
and through the dirty windows of the windblown car
they see that they move forward toward a grey snowless sky between walls of grey and greyish-brown
buildings with brick fronts, with broken glass between
the brick fronts and debris in the streets.

"Where are we?" Chris says, and Phaedrus says, "I
don't know," and he really doesn't, his mind is all but
gone. He is lost, drifting through the grey streets.

"Where are we going?" says Phaedrus.

"To the bunk-bedders," says Chris.

"Where are *they*?" asks Phaedrus.

"I don't know," says Chris. "Maybe if we just keep going we'll see them."

And so the two drive and drive through the endless streets looking for the bunk-bedders. Phaedrus wants to stop and put his head on the steering wheel and just rest. The soot and the grey have penetrated his eyes and all but blotted cognizance from his brain. One street sign is like another. One grey-brown building is like the next. On and on they drive, looking for the bunk-bedders. But the bunk-bedders, Phaedrus knows, he will never find.

Chris begins to realize slowly and by degrees that something is strange, that the person guiding the car is no longer really guiding it, that the captain is dead and the car is pilotless and he doesn't know this but only feels it and says stop and Phaedrus stops.

A car behind honks, but Phaedrus does not move. Other cars honk, and then others, and Chris in panic says, "GO!" and Phaedrus slowly with agony pushes his foot on the clutch and puts the car in gear. Slowly, in dream-motion, the car moves in low through the streets.

"Where do we live?" Phaedrus asks a frightened Chris.

Chris remembers an address, but doesn't know how to get there, but reasons that if he asks enough people he will find the way and so says, "Stop the car," and gets out and asks directions and leads a demented Phaedrus through the endless walls of brick and broken glass.

Hours later they arrive and the mother is furious that they are so late. She cannot understand why they have not found the bunk-bedders. Chris says, "We looked everywhere," but looks at Phaedrus with a quick glance of fright, of terror at something unknown. That, for Chris, is where it started.

It won't happen again. . . .

I think what I'll do is head down for San Francisco, and put Chris on a bus for home, and then sell the cycle and check in at a hospital . . . or that last seems so pointless . . . I don't know what I'll do.

The trip won't have been entirely wasted. At least he'll have some good memories of me as he grows up. That takes away some of the anxiety a little. That's a good thought to hold on to. I'll hold on to that.

Meanwhile, just continue on a normal trip and hope something improves. Don't throw anything away. Never, never throw anything away.

Cold out! Feels like winter! Where are we, that it should get this cold? We must be at a high altitude. I look out of the sleeping bag and this time see frost on the motorcycle. On the chrome of the gas tank it's sparkling in the early sunlight. On the black frame where the sunlight hits it it's partly turned to beads of water that will soon run down to the wheel. It's too cold to lie around.

I remember the dust under the pine needles and put my boots on carefully to avoid stirring it up. At the motorcycle I unpack everything, get out the long underwear and put it on, then clothes, then sweater, then jacket. I'm still cold.

I step through the spongy dust onto the dirt road that has brought us here and sprint down it through the pines for a hundred feet or so, then settle down to an even run and then finally stop. That feels better. Not a sound. The frost is in little patches on the road too, but melting and dark wet tan between the patches where the early sun's rays strike it. It's so white and lacy and untouched. It's on the trees too. I walk back softly down the road as if not to disturb the sunrise. Early autumn feeling.

Chris is still asleep and we won't be able to go anywhere until the air warms up. Good time to get the cycle tuned. I work loose the knob on the side cover over the air filter, and underneath the filter withdraw a worn and dirty roll of the field tools. My hands are stiff with the cold and the backs of them are wrinkled. Those wrinkles aren't from the cold though. At forty that's old age coming on. I lay the roll on the seat and spread it open . . . there they are . . . like seeing old friends again.

I hear Chris, glance over the seat and see that he's stirring but doesn't get up. He's evidently just rolling in his sleep. After a while the sun gets warmer and my hands aren't as stiff as they were.

I was going to talk about some of the lore of cycle repair, the hundreds of things you learn as you go along, which enrich what you're doing not only practically but esthetically. But that seems too trivial now, though I shouldn't say that.

But now I want to shift into another direction, which completes *his* story. I never really completed it because I didn't think it would be necessary. But now I think it would be a good time to do that in what time remains.

The metal of these wrenches is so cold it hurts the hands. But it's a good hurt. It's real, not imaginary, and it's here, absolutely, in my hand.

*   *   *

. . . When you travel a path and note that another path breaks away to one side at, say, a 30-degree angle, and then later another path branches away to the same side at a broader angle, say 45 degrees, and another path later at 90 degrees, you begin to understand that there's some point over there that all the paths lead to and that a lot of people have found it worthwhile to go that way, and you begin to wonder out of curiosity if perhaps that isn't the way you should go too.

In his pursuit of a concept of Quality, Phaedrus kept seeing again and again little paths all leading toward some point off to one side. He thought he already knew about the general area they led to, ancient Greece, but now he wondered if he had overlooked something there.

He had asked Sarah, who long before had come by with her watering pot and put the idea of Quality in his head, where in English literature quality, as a subject, was taught.

"Good heavens, I don't know, I'm not an English scholar," she had said. "I'm a classics scholar. My field is Greek."

"Is quality a part of Greek thought?" he had asked.

"Quality is *every* part of Greek thought," she had said, and he had thought about this. Sometimes under her old-ladyish way of speaking he thought he detected a secret canniness, as though like a Delphic oracle she said things with hidden meanings, but he could never be sure.

Ancient Greece. Strange that for them Quality should be everything while today it sounds odd to even say quality is real. What unseen changes could have taken place?

A second path to ancient Greece was indicated by the sudden way the whole question, What is quality?, had been jolted into systematic philosophy. He had thought he was done with that field. But "quality" had opened it all up again.

Systematic philosophy is Greek. The ancient Greeks invented it and, in so doing, put their permanent stamp on it. Whitehead's statement that all philosophy is nothing but "footnotes to Plato" can be well supported. The confusion about the reality of Quality had to start back there sometime.

A third path appeared when he decided to move on from Bozeman toward the Ph.D. degree he needed to continue University teaching. He wanted to pursue the inquiry into the meaning of Quality that his English teaching had started. But where? And in which discipline?

It was apparent that the term "Quality" was not within any one discipline unless that discipline was philosophy. And he knew from his experience with philosophy that further study there was unlikely to uncover anything concerning an apparently mystic term in English composition.

He became more and more aware of the possibility that there was no program available where he might study Quality in terms resembling those in which he understood it. Quality lay not only outside any academic discipline, it lay outside the grasp of the methods of the entire Church of Reason. It would take quite a University to accept a doctoral thesis in which the candidate refused to define his central term.

He looked through the catalogs for a long time before he discovered what he hoped he was looking for.

There *was* one University, the University of Chicago, where there existed an interdisciplinary program in "Analysis of Ideas and Study of Methods." The examining committee included a professor of English, a professor of philosophy, a professor of Chinese, and the Chairman, who was a professor of ancient Greek! That one rang bells.

On the machine now everything is done except the oil change. I wake Chris and we pack and go. He's still sleepy but the cold air on the road wakes him up.

The piney road goes upward, and there's not so much traffic this morning. The rocks among the pines are dark and volcanic. I wonder if that was volcanic dust we slept in. Is there such a thing as volcanic dust? Chris says he's hungry and I am too.

At La Pine we stop. I tell Chris to order me ham and eggs for breakfast while I stay outside to change the oil.

At a filling station next to the restaurant I pick up a quart of oil, and in a gravelly lot back of the restaurant remove the drain plug, let the oil drain, replace the plug, add the new oil, and when I'm done the new oil on the dipstick shines in the sunlight almost as clear and colorless as water. Ahhhhh!

I repack the wrench, enter the restaurant and see Chris and, on the table, my breakfast. I head into the washroom, clean up and return.

"Am I hungry!" he says.

"It was a cold night," I say. "We burned up a lot of food just staying alive."

The eggs are good. The ham too. Chris talks about the dream and how it frightened him and then that's done with. He looks as though he's about to ask a question, then doesn't, then stares out the window into the pines for a while, then comes back with it.

"Dad?"

"What?"

"Why are we doing this?"

"What?"

"Just riding all the time."

"Just to see the country . . . vacation."

The answer doesn't seem to satisfy him. But he can't seem to say what's wrong with it.

A sudden despair wave hits, like that at dawn. I *lie* to him. That's what's wrong.

"We just keep going and going," he says.

"Sure. What would you rather do?"

He has no answer.

I don't either.

On the road an answer comes that we're doing the highest-Quality thing I can think of right now, but that wouldn't satisfy him any more than what I told him. I don't know what else I could have said. Sooner or later, before we say good-bye, if that's how it goes, we'll have to do some talking. Shielding him like this from the past may be doing him more harm than good. He'll have to hear about Phaedrus, although there's much he can never know. Particularly the end.

Phaedrus arrived at the University of Chicago already in a world of thought so different from the one you or I understand, it would be difficult to relate, even if I fully remembered everything. I know that the acting chairman admitted him during the Chairman's absence on the basis of his teaching experience and apparent ability to converse intelligently. What he actually said is lost. Afterward he waited for a number of weeks for the Chairman to return in hopes of obtaining a scholarship, but when the Chairman did appear an interview took place which consisted essentially of one question and no answer.

The Chairman said, "What is your substantive field?"

Phaedrus said, "English composition."

The Chairman bellowed, "That is a methodological field!" And for all practical purposes that was the end of the interview. After some inconsequential conversation Phaedrus stumbled, hesitated and excused himself, then went back to the mountains. This was the characteristic of his that had failed him out of the University before. He had gotten stuck on a question and hadn't been able to think about anything else, while the classes moved on without him. This time, however, he had all summer to think about why his field should be substantive or methodological, and all that summer that is what he did.

In the forests near the timberline he ate Swiss cheese, slept on pine-bough beds, drank mountain stream water and thought about Quality and substantive and methodological fields.

Substance doesn't change. Method contains no permanence. Substance relates to the form of the atom. Method relates to what the atom does. In technical composition a similar distinction exists between physical description and functional description. A complex assembly is best described first in terms of its substances: its subassemblies and parts. Then, next, it is described in terms of its methods: its functions as they occur in sequence. If you confuse physical and functional description, substance and method, you get all tangled up and so does the reader.

But to apply these classifications to a whole field of knowledge such as English composition seemed arbitrary and impractical. No academic discipline is without both substantive and methodological aspects. And Quality had no connection that he could see with either one of them. Quality isn't a substance. Neither is it a method. It's outside of both. If one builds a house using the plumb-line and spirit-level methods he does so because a straight vertical wall is less likely to collapse and thus has higher Quality than a crooked one. Quality isn't method. It's the goal toward which method is aimed.

"Substance" and "substantive" really corresponded to "object" and "objectivity," which he'd rejected in order to arrive at a nondualistic concept of Quality. When everything is divided up into substance and method, just as when everything's divided up into subject and object, there's really no room for Quality at all. His thesis could not be a part of a substantive field, because to accept a split into substantive and methodological was to deny the existence of Quality. If Quality was going to stay, the concept of substance and method would have to go. That would mean a quarrel with the committee, something he had no desire for at all. But he was angry that they should destroy the entire meaning of what he was saying with the very first question. Substantive field? What kind of Procrustean bed were they trying to shove him into? he wondered.

He decided to examine more closely the background of the committee and did some library digging for this purpose. He felt this committee was off into some entirely alien pattern of thought. He didn't see where this pattern and the large pattern of his own thought joined together.

He was especially disturbed by the quality of the explanations of the committee's purpose. They seemed extremely confusing. The entire description of the committee's work was a strange pattern of ordinary enough words put together in a most unordinary way, so that the explanation seemed far more complex than the thing he was trying to have explained. This wasn't the bells ringing he'd heard before.

He studied everything he could find that the Chairman had written and here again was found the strange pattern of language seen in the confusing description of the committee. It was a puzzling style because it was completely different from what he'd seen of the Chairman himself. The Chairman, in a brief interview, had impressed him with great quickness of mind, and an equally swift temper. And yet here was one of the most ambiguous, inscrutable styles Phaedrus had ever read. Here were encyclopedic sentences that left subject and predicate completely out of shouting distance. Parenthetic elements were unexplainably inserted inside other parenthetic elements, equally unexplainably inserted into sentences whose relevance to the preceding sentences in the reader's mind was dead and buried and decayed long before the arrival of the period.

But most remarkable of all were the wondrous and unexplained proliferations of abstract categories that seemed freighted with special meanings that never got stated and whose content could only be guessed at; these piled one after another so fast and so close that

Phaedrus knew he had no possible way of understanding what was before him, much less take issue with it.

At first Phaedrus presumed the reason for the difficulty was that all this was over his head. The articles assumed a certain basic learning which he didn't have. Then, however, he noticed that some of the articles were written for audiences that couldn't possibly have this background, and this hypothesis was weakened.

His second hypothesis was that the Chairman was a "technician," a phrase he used for a writer so deeply involved in his field that he'd lost the ability to communicate with people outside. But if this were so, why was the committee given such a general, nontechnical title as "Analysis of Ideas and Study of Methods"? And the Chairman didn't have the personality of a technician. So that hypothesis was weak too.

In time, Phaedrus abandoned the labor of pounding his head against the Chairman's rhetoric and tried to discover more about the background of the committee, hoping *that* would explain what this was all about. This, it turned out, was the correct approach. He began to see what his trouble was.

The Chairman's statements were guarded—guarded by enormous, labyrinthine fortifications that went on and on with such complexity and massiveness it was almost impossible to discover what in the world it was inside them he was guarding. The inscrutability of all this was the kind of inscrutability you have when you suddenly enter a room where a furious argument has just ended. Everyone is quiet. No one is talking.

I have one tiny fragment of Phaedrus standing in the stone corridor of a building, evidently within the University of Chicago, addressing the assistant chairman of the committee, like a detective at the end of a movie, saying: "In your description of the committee, you have omitted one important name."

"Yes?" says the assistant chairman.

"Yes," says Phaedrus omnisciently, ". . . Aristotle . . ."

The assistant chairman is shocked for a moment, then, almost like a culprit who has been discovered but feels no guilt, laughs loud and long.

"Oh, I see," he says. "You didn't know . . . anything about . . ." Then he thinks better of what he is going to say and decides not to say anything more.

We arrive at the turnoff to Crater Lake and go up a neat road into the National Park—clean, tidy and preserved. It really shouldn't be any other way, but this doesn't win any prizes for Quality either. It turns it into a museum. This is how it was before the white man came— beautiful lava flows, and scrawny trees, and not a beer can anywhere—but now that the white man is here, it looks fake. Maybe the National Park Service should set just one pile of beer cans in the middle of all that lava and then it would come to life. The absence of beer cans is distracting.

At the lake we stop and stretch and mingle affably with the small crowd of tourists holding cameras and children yelling, "Don't go too close!" and see cars and campers with all different license plates, and see the Crater Lake with a feeling of "Well, there it is," just as the pictures show. I watch the other tourists, all of whom seem to have out-of-place looks too. I have no resentment at all this, just a feeling that it's all unreal and that the quality of the lake is smothered by the fact that it's so pointed to. You point to something as having Quality and the Quality tends to go away. Quality is what you see out of the corner of your eye, and so I look at the lake below but feel the peculiar quality from the chill, almost frigid sunlight behind me, and the almost motionless wind.

"Why did we come here?" Chris says.

"To see the lake."

He doesn't like this. He senses falseness and frowns deep, trying to find the right question to expose it. "I just hate this," he says.

A tourist lady looks at him with surprise, then resentment.

"Well, what can we do, Chris?" I ask. "We just have to keep going until we find out what's wrong or find out why we don't know what's wrong. Do you see that?"

He doesn't answer. The lady pretends not to be listening, but her motionlessness reveals that she is. We walk toward the motorcycle, and I try to think of something, but nothing comes. I see he's crying a little and now looks away to prevent me from seeing it.

We wind down out of the park to the south.

I said the assistant chairman for the Committee on Analysis of Ideas and Study of Methods was shocked. What he was so shocked about was that Phaedrus didn't know he was at the locus of what is probably the most famous academic controversy of the century, what a California university president described as the last attempt in history to change the course of an entire university.

Phaedrus' reading turned up a brief history of that famous revolt against empirical education that had taken place in the early thirties. The Committee on Analysis of Ideas and Study of Methods was a vestige of that attempt. The leaders of the revolt were Robert Maynard Hutchins, who had become president of the University of Chicago; Mortimer Adler, whose work on the psychological background of the law of evidence was somewhat similar to work being done at Yale by Hutchins; Scott Buchanan, a philosopher and mathematician; and most important of all for Phaedrus, the present chairman of the committee, who was then a Columbia University Spinozist and medievalist.

Adler's study of evidence, cross-fertilized by a reading of classics of the Western World, resulted in a conviction that human wisdom had advanced relatively little in recent times. He consistently harked back to St. Thomas Aquinas, who had taken Plato and Aristotle and made them part of his medieval synthesis of Greek philosophy and Christian faith. The work of Aquinas and of the Greeks, as interpreted by Aquinas, was to Adler the capstone of the Western intellectual heritage. Therefore they provided a measuring rod for anyone seeking the good books.

In the Aristotelian tradition as interpreted by the medieval scholastics, man is counted a rational animal, capable of seeking and defining the good life and achieving it. When this "first principle" about the nature of man was accepted by the president of the University of Chicago, it was inevitable that it would have educational repercussions. The famous University of Chicago Great Books program and the reorganization of the University structure along Aristotelian lines and the establishment of the "College," in which a reading of classics was initiated in fifteen-year-old students, were some of the results.

Hutchins had rejected the idea that an empirical scientific education could automatically produce a "good" education. Science is "value free." The inability of science to grasp Quality, as an object of inquiry, makes it impossible for science to provide a scale of values.

Adler and Hutchins were concerned fundamentally with the "oughts" of life, with values, with Quality and with the foundations of Quality in theoretical philosophy. Thus they had apparently been traveling in the same direction as Phaedrus but had somehow ended with Aristotle and stopped there.

There was a clash.

Even those who were willing to admit Hutchins' preoccupation with Quality were unwilling to grant the final authority to the Aristotelian tradition to define values. They insisted that no values can be fixed, and that a valid modern philosophy need not reckon with ideas as they are expressed in the books of ancient and medieval times. The whole business seemed to many of them merely a new and pretentious jargon of weasel concepts.

Phaedrus didn't know quite what to make of this clash. But it certainly seemed to be close to the area he wished to work in. He also felt that no values can be fixed but that this is no reason why values should be ignored or that values do not exist as reality. He also felt antagonistic to the Aristotelian tradition as a definer of values, but he didn't feel this tradition should be left unreckoned with. The answer to all this was somehow deeply enmeshed in it and he wanted to know more.

Of the four who had created such a furor, the present chairman of the committee was the only one now left. Perhaps because of this reduction in rank, perhaps for other reasons, his reputation among persons Phaedrus talked to wasn't one of geniality. His geniality was confirmed by none and sharply refuted by two, one the head of a major University department who described him as a "holy terror" and another who held a graduate degree in philosophy from the University of Chicago who said the chairman was well known for graduating only carbon copies of himself. Neither of these advisers was by nature vindictive and Phaedrus felt what they said was true. This was further confirmed by a discovery made at the department office. He wanted to talk to two graduates of the committee to find out more of what it was about, and had been told that the committee had granted only two Ph.D.'s in its history. Apparently to find room in the sun for a reality of Quality he would have to fight and overcome the head of his own committee, whose Aristotelian outlook made it impossible even to get started and whose temperament appeared to be extremely intolerant of opposing ideas. It all added up to a very gloomy picture.

He then sat down and penned, to the Chairman of the Committee on Analysis of Ideas and Study of Methods at the University of Chicago, a letter which can only be described as a provocation to dismissal, in which the writer refuses to skulk quietly out the back door but instead creates a scene of such proportions the opposition is forced to throw him out the *front* door, thus giving weight to the provocation it didn't formerly have. Afterward he picks himself up out of the street and, after making sure the door is completely closed, shakes his fist at it, dusts himself off and says, "Oh well, I *tried*"; and in this way absolves his conscience.

Phaedrus' provocation informed the Chairman that his substantive field was now philosophy, not English composition. However, he said, the division of study into substantive and methodological fields was an outgrowth of the Aristotelian dichotomy of form and substance, which nondualists had little use for, the two being identical.

He said he wasn't sure, but the thesis on Quality appeared to turn into an anti-Aristotelian thesis. If this was true he had chosen an appropriate place to present it. Great Universities proceeded in a Hegelian fashion and any school which could not accept a thesis contradicting its fundamental tenets was in a rut. Thus, Phaedrus claimed, was the thesis the University of Chicago was waiting for.

He admitted the claim was grandiose and that value judgments were actually impossible for him to make since no person could be an impartial judge of his own cause. But if someone else were to produce a thesis which purported to be a major breakthrough between Eastern and Western philosophy, between religious mysticism and scientific positivism, he would think it of major historic importance, a thesis which would place the University miles ahead. In any event, he said, no one was really accepted in Chicago until he'd rubbed someone out. It was time Aristotle got his.

Just outrageous.

And not just provocation to dismissal either. What comes through even more strongly is megalomania, delusions of grandeur, of complete loss of ability to understand the effect of what he was saying on others.

He had become so caught up in his own world of Quality metaphysics he couldn't see outside it anymore, and since no one else understood this world, he was already done for.

I think he must have felt at the time that what he was saying was true and it didn't matter if his manner or presentation was outrageous or not. There was so much to it he didn't have time for prettying it up. If the University of Chicago was interested in the esthetics of what he was saying rather than the rational content, they were failing their fundamental purpose as a University.

This was it. He really *believed*. It wasn't just another interesting idea to be tested by existing rational methods. It was a modification of the existing rational methods themselves. Normally when you have a new idea to present in an academic environment you're supposed to be objective and disinterested in it. But this idea of Quality took issue with that very supposition—of objectivity and disinterestedness. These were mannerisms appropriate only to dualistic reason. Dualistic excellence is achieved by objectivity, but creative excellence is *not*.

He had the faith that he had solved a huge riddle of the universe, cut a Gordian knot of dualistic thought with one word, Quality, and he wasn't about to let anyone tie that word down again. And in believing, he couldn't see how outrageously megalomaniacal his words sounded to others. Or if he saw it, didn't care. What he said was megalomaniacal, but suppose it was *true*? If he was wrong, who would care? But suppose he was *right*? To be right and throw it away in order to please the predilections of his teachers, *that* would be the monstrosity!

And so he just did not care how he sounded to others. It was a totally fanatic thing. He lived in a solitary universe of discourse in those days. No one understood him. And the more people showed how they failed to understand him and disliked what they did understand, the more fanatic and unlikable he became.

His provocation to dismissal was given an expected reception. Since his substantive field was philosophy, he should apply to the philosophy department, not the committee.

Phaedrus dutifully did this, then he and his family loaded their car and trailer with all they owned and said good-bye to their friends and were about to start. Just as he locked the doors of the house for the last time the mailman appeared with a letter. It was from the University of Chicago. It said he was not admitted there. Nothing more.

Obviously the Chairman of the Committee on Analysis of Ideas and Study of Methods had influenced the decision.

Phaedrus borrowed some stationery from the neighbors and wrote back to the Chairman that since he had *already been admitted* to the Committee on Analysis of Ideas and Study of Methods he would have to *remain* there. This was a rather legalistic maneuver, but Phaedrus by this time had developed a kind of combative canniness. This deviousness, the quick shuffle out the philosophy door seemed to indicate that the Chairman for some reason was *unable* to throw him out the front door of the committee, even with that outrageous letter in hand, and that gave Phaedrus some confidence. No side doors, please. They were going to have to throw him out the front door or not at all.

Maybe they wouldn't be able to. Good. He wanted this thesis not to owe anyone anything.

We travel down the eastern shore of Klamath Lake on a three-lane highway that contains a lot of nineteen-twenties feeling. That's when these three-laners were all made. We pull in for lunch at a roadhouse which belongs to this era too. Wooden frame badly in need of paint, neon beer signs in the window, gravel and engine drippings for a front lawn.

Inside, the toilet seat is cracked and the washbowl is covered with grease streaks, but on my way back to our booth I take a second look at the owner behind the bar. A nineteen-twenties face. Uncomplicated, uncool and unbowed. This is his castle. We're his guests. And if we don't like his hamburgers we'd better shut up.

When they arrive, the hamburgers, with giant raw onions, are tasty and the bottle beer is fine. A whole meal for a lot less than you'd pay at one of those old-ladies places with plastic flowers in the window. As we eat I see on the map we've taken a wrong turn away back and could have gotten to the ocean much quicker by another route. It's hot now, a West Coast sticky hotness which after the Western Desert hotness is very depressing. Really, this is just transported East, all of this scene, and I'd like to get to the ocean where it's cool as soon as possible.

I think about this all around the southern shore of Klamath Lake. Sticky hotness and nineteen-twenties funk. . . . That was the feeling of Chicago that summer.

When Phaedrus and his family arrived in Chicago, he took up residence near the University and, since he had no scholarship, began full-time teaching of rhetoric at the University of Illinois, which was then downtown at Navy Pier, sticking out into the lake, funky and hot.

Classes were different from those in Montana. The top high-school students had been skimmed off to the Champaign and Urbana campuses and almost all the students he taught were a solid monotonous C. When their papers were judged in class for Quality it was hard to distinguish among them. Phaedrus, in other circumstances, probably would have invented something to get around this, but now this was just bread-and-butter work for which he couldn't spare creative energy. His interest lay to the south at the other University.

He entered the University of Chicago registration lineup, announced his name to the registering Professor of Philosophy and noticed a slight setting of the eyes. The Professor of Philosophy said, Oh, yes, the Chairman had asked that he be registered in an Ideas and Methods course which the Chairman himself was teaching, and give him the schedule of the course. Phaedrus noted that the time set for the class conflicted with his schedule at Navy Pier and chose instead another one, Ideas and Methods 251, Rhetoric. Since rhetoric was his own field, he felt a little more at home here. And the lecturer wasn't the Chairman. The lecturer was the Professor of Philosophy now registering him. The Professor of Philosophy's eyes, formerly set, now became wide.

Phaedrus returned to his teaching at Navy Pier and his reading for his first class. It was now absolutely necessary that he study as he had never studied before to learn the thought of Classic Greece in general and of one Classic Greek in particular—Aristotle.

Of all the thousands of students at the University of Chicago who had studied the ancient classics it's doubtful that there was ever a more dedicated one. The main struggle of the University's Great Books program was against the modern belief that the classics had nothing of any real importance to say to a twentieth-century society. To be sure, the majority of students taking the courses must have played the game of nice manners with their teachers, and accepted, for purposes of understanding, the prerequisite belief that the ancients had something meaningful to say. But Phaedrus, playing no games at all, didn't just *accept* this idea. He passionately and fanatically *knew* it. He came to *hate* them vehemently, and to assail them with every kind of invective he could think of, not because they were irrelevant but for exactly the opposite reason. The more he studied, the more convinced he became that no one had yet told the damage to this world that had resulted from our unconscious acceptance of their thought.

Around the southern shore of Klamath Lake we pass through some suburban-type development, and then leave the lake to the west, toward the coast. The road goes up now into the forests of huge trees not at all like the rain-starved forests we've been through. Huge Douglas firs are on either side of the road. On the cycle we can look up along their trunks, straight up, for hundreds of feet as we pass between them. Chris wants to stop and walk among them and so we stop.

While he goes for a walk I lean my back as carefully as possible against a big slab of Douglas fir bark and look up and try to remember. . . .

* * *

The details of what he learned are lost now, but from events that occurred later I know he absorbed tremendous quantities of information. He was capable of doing this on a near-photographic basis. To understand how he arrived at his condemnation of the Classic Greeks it's necessary to review in summary form the "mythos over logos" argument, which is well known to scholars of Greek and is often a cause of fascination with that area of study.

The term *logos,* the root word of "logic," refers to the sum total of our rational understanding of the world. *Mythos* is the sum total of the early historic and prehistoric myths which preceded the logos. The mythos includes not only the Greek myths but the Old Testament, the Vedic Hymns and the early legends of all cultures which have contributed to our present world understanding. The mythos-over-logos argument states that our rationality is shaped by these legends, that our knowledge today is in relation to these legends as a tree is in relation to the little shrub it once was. One can gain great insights into the complex overall structure of the tree by studying the much simpler shape of the shrub. There's no difference in kind or even difference in identity, only a difference in size.

Thus, in cultures whose ancestry includes ancient Greece, one invariably finds a strong subject-object differentiation because the grammar of the old Greek mythos presumed a sharp natural division of subjects and predicates. In cultures such as the Chinese, where subject-predicate relationships are not rigidly defined by grammar, one finds a corresponding absence of rigid subject-object philosophy. One finds that in the Judeo-Christian culture, in which the Old Testament "Word" had an intrinsic sacredness of its own, men are willing to sacrifice and live by and die for words. In this culture, a court of law can ask a witness to tell "the truth, the whole truth and nothing but the truth, so help me God," and expect the truth to be told. But one can transport this court to India, as did the British, with no real success on the matter of perjury because the Indian mythos is different and this sacredness of words is not felt in the same way. Similar problems have occurred in this country among minority groups with different cultural backgrounds. There are endless examples of how mythos differences direct behavior differences and they're all fascinating.

The mythos-over-logos argument points to the fact that each child is born as ignorant as any caveman. What keeps the world from reverting to the Neanderthal with each generation is the continuing, ongoing mythos, transformed into logos but still mythos, the huge body of common knowledge that unites our minds as cells are united in the body of man. To feel that one is not so united, that one can accept or discard this mythos as one pleases, is not to understand what the mythos is.

There is only one kind of person, Phaedrus said, who accepts or rejects the mythos in which he lives. And the definition of that person, when he has rejected the mythos, Phaedrus said, is "insane." To go outside the mythos is to become insane. . . .

My God, that just came to me now. I never knew that before.

He knew! He must have known what was about to happen. It's starting to open up.

You have all these fragments, like pieces of a puzzle, and you can place them together into large groups, but the groups don't go together no matter how you try, and then suddenly you get one fragment and it fits two different groups and then suddenly the two great groups are one. The relation of the mythos to insanity. That's a key fragment. I doubt whether anyone ever said that before. Insanity is the *terra incognita* surrounding the mythos. And he knew! He knew the Quality he talked about lay outside the mythos.

Now it comes! Because Quality is the *generator* of the mythos. That's it. That's what he meant when he said, "Quality is the continuing stimulus which causes us to create the world in which we live. All of it. Every last bit of it." Religion isn't invented by man. Men are invented by religion. Men invented *responses* to Quality, and among these responses is an understanding of what they themselves are. You know something and then the Quality stimulus hits and then you try to define the Quality stimulus, but to define it all you've got to work with is what you know. So your definition is made up of what you know. It's an analogue to what you already know. It *has* to be. It can't be anything else. And the mythos grows this way. By analogies to what is known before. The mythos is a building of analogues upon analogues upon analogues. These fill the boxcars of the train of consciousness. The mythos is the whole train of collective consciousness of all communicating mankind. Every last bit of it. The Quality is the track that directs the train. What is outside the train, to either side—that is the *terra incognita* of the insane. He knew that to understand Quality he would have to leave the mythos. That's why he felt that slippage. He knew something was about to happen.

I see Chris running through the trees now. He looks relaxed and happy. He shows me a piece of bark and asks if he can save it as a souvenir. I haven't been fond of loading the cycle with these bits and pieces he finds and will probably throw away when he gets home, but this time say okay anyway.

After a few minutes the road reaches a summit and then drops steeply into a valley that becomes more exquisite as we descend. I never thought I would call a valley that—exquisite—but there's something about this whole coastal country so different from any other mountainous region in America that it brings out the word. Here, a little farther south, is where all our good wine comes from. The hills are somehow tucked and folded differently—exquisitely. The road twists and banks and curlicues and descends and we and the cycle smoothly roll with it, following it in a separate grace of our own, almost touching the waxen leaves of shrubs and overhanging boughs of trees. The firs and rocks of the higher country are behind us now and around us are soft hills and vines and purple and red flowers, fragrance mixed with woodsmoke up from the distant fog along the valley floor and from beyond that, unseen—a vague scent of ocean. . . .

. . . How can I love all this so much and be insane? . . .

. . . I don't *believe* it!

The mythos. The mythos is insane. That's what he

believed. The mythos that says the forms of this world are real but the Quality of this world is unreal, that is insane!

And in Aristotle and the ancient Greeks he believed he had found the villains who had so shaped the mythos as to cause us to accept this insanity as reality.

That. That now. That ties it all together. It feels relieving when that happens. It's so hard sometimes to conjure all this up, a strange sort of exhaustion follows. Sometimes I think I'm just making it up myself. Sometimes I'm not sure. And sometimes I know I'm not. But the mythos and insanity, and the centrality of this—this I'm sure is from *him*.

When we're through the folded hills we come to Medford and a freeway leading to Grants Pass and it's almost evening. A heavy head wind keeps us just up with traffic on upgrades, even with the throttle wide open. Coming into Grants Pass we hear a frightening, loud, clanking noise and stop to discover that the chain guard has become caught in the chain somehow and now is all torn up. Not too serious, but enough to lay us up for a while to get it replaced. Foolish to replace it, perhaps, when the cycle will be sold in a few days.

Grants Pass looks like a big enough town to have a motorcycle place open the next morning and when we arrive I look for a motel.

We haven't seen a bed since Bozeman, Montana.

We find one with color TV, heated swimming pool, a coffee maker for the next morning, soap, white towels, a shower all tiled and clean beds.

We lie down on the clean beds and Chris just

bounces on his for a while. Bouncing on beds, I remember from childhood, is a great depression reliever.

Tomorrow, somehow, all this can be worked out, maybe. Not now. Chris goes down for a heated swim while I lie quietly on the clean bed and put everything out of mind.

29

※

In the process of taking stuff out of the saddlebags and cramming it back ever since Bozeman, and doing the same with the backpacks, we've acquired some exceptionally beat-up gear. Spread out all over the floor in the morning light it looks a mess. The plastic bag with the oily stuff in it has broken and oil has gotten onto the roll of toilet paper. The clothes have been so squashed they look as if they have permanent, built-in wrinkles. The soft metal tube of sunburn ointment has burst, leaving white crud all over the machete scabbard and a fragrant smell everywhere. The tube of ignition grease has burst too. What a mess. In my shirt-pocket notebook I write down: "Buy tackle box for squeezed stuff" and then add "Do laundry." Then, "Buy toenail scissors, sunburn cream, ignition grease, chain guard, toilet paper." This is a lot of things to do before checkout time, so I wake up Chris and tell him to get up. We have to do the laundry.

At the Laundromat I instruct Chris on how to operate the drier, start the washing machines and take off for the other items.

I get everything but a chain guard. The parts man says they don't have one and don't expect to get one. I think about riding without the chain guard for what little time is left but that will throw crud all over and could be dangerous. Also, I don't want to do things with that presumption. That commits me to it.

Down the street I find a welder's sign and enter.

Cleanest welding place I've ever seen. Great high trees and deep grass line an open space in back, giving a kind of village-smithy appearance. All the tools are hung up with care, everything tidy, but no one is home. I'll come back later.

I wheel back and stop for Chris, check the laundry he's put in the drier and putt along through the cheerful streets looking for a restaurant. Traffic everywhere, alert, well-maintained cars, most of them. West Coast. Hazy clean sunlight of a town out of the range of the coal vendors.

At the edge of town we find a restaurant and sit and wait at a red and white tableclothed table. Chris unfolds a copy of *Cycle News,* which I bought at the cycle shop, and reads out loud who has won all the races, and an item about cross-country cycling. The waitress looks at him, a little curiously, and then at me, then at my cycle boots, then jots down our order. She goes back into the kitchen and comes out again and looks at us. I guess that she's paying so much attention to us because we're alone here. While we wait she puts some coins in the jukebox and when breakfast comes—waffles, syrup and sausages, ah—we have music with it. Chris and I talk about what he sees in *Cycle News* and we are talking over the noise of the record in the relaxed way people talk who have been many days on the road together and out of the corner of my eye I see that this is watched with a steady gaze. After a while Chris has to ask me some questions a second time because that gaze kind of beats on me, and it's hard to think of what he's saying. The record is a country western about a truck driver. . . . I finish the conversation with Chris.

As we leave and go out and start up the cycle, there she is in the door watching us. Lonely. She probably doesn't understand that with a look like that she isn't going to be lonely long. I kick the starter and gun the engine too hard, frustrated by something, and as we ride for the welder again, it takes a while to snap out.

The welder is in, an old man in his sixties or seventies, and he looks at me disdainfully—a complete reversal from the waitress. I explain about the chain guard and after a while he says, "I'm not taking it off for you. You'll have to take it off."

I do this and show it to him, and he says, "It's full of grease."

I find a stick out in back under the spreading chestnut tree and scrape all the grease into a trash barrel. From a distance he says, "There's some solvent in that pan over there." I see the flat pan and get out the remaining grease with some leaves and the solvent.

When I show it to him he nods and slowly goes over and sets the regulators for his gas torch. Then he looks at the tip and selects another one. Absolutely no hurry. He picks up a steel filler rod and I wonder if he's actually going to try to *weld* that thin metal. Sheet metal I don't weld. I braze it with a brass rod. When I try to weld it I punch holes in it and then have to patch them up with huge blobs of filler rod. "Aren't you going to braze it?" I ask.

"No," he says. Talkative fellow.

He sparks the torch, and sets a tiny little blue flame and then, it's hard to describe, actually dances the torch and the rod in separate little rhythms over the thin sheet metal, the whole spot a uniform luminous orange-yellow, dropping the torch and filler rod down at the exact right moment and then removing them. No holes. You can hardly see the weld. "That's beautiful," I say.

"One dollar," he says, without smiling. Then I catch a funny quizzical look within his glance. Does he wonder if he's overcharged? No, something else . . . lonely, same as the waitress. Probably he thinks I'm bullshitting him. Who appreciates work like this anymore?

We're packed and out of the motel at just about checkout time and are soon into the coastal redwood forest, across out of Oregon into California. The traffic is so heavy we don't have time to look up. It's turning cold and grey and we stop and put on sweaters and jackets. It's still cold, somewhere in the low fifties, and we think winter thoughts.

Lonely people back in town. I saw it in the supermarket and at the Laundromat and when we checked out from the motel. These pickup campers through the redwoods, full of lonely retired people looking at trees on their way to look at the ocean. You catch it in the first fraction of a glance from a new face—that searching look—then it's gone.

We see much more of this loneliness now. It's paradoxical that where people are the most closely crowded, in the big coastal cities in the East and West, the loneliness is the greatest. Back where people were so spread out in western Oregon and Idaho and Montana and the Dakotas you'd think the loneliness would have been greater, but we didn't see it so much.

The explanation, I suppose, is that the physical distance between people has nothing to do with loneliness. It's psychic distance, and in Montana and Idaho the physical distances are big but the psychic distances between people are small, and here it's reversed.

It's the primary America we're in. It hit the night before last in Prineville Junction and it's been with us ever since. There's this primary America of freeways and jet flights and TV and movie spectaculars. And people caught up in this primary America seem to go through huge portions of their lives without much consciousness of what's immediately around them. The media have convinced them that what's right around them is unimportant. And that's why they're lonely. You see it in their faces. First the little flicker of searching, and then when they look at you, you're just a kind of an object. You don't count. You're not what they're looking for. You're not on TV.

But in the secondary America we've been through, of back roads, and Chinaman's ditches, and Appaloosa horses, and sweeping mountain ranges, and meditative thoughts, and kids with pinecones and bumblebees and open sky above us mile after mile after mile, all through that, what was real, what was *around* us dominated. And so there wasn't much feeling of loneliness. That's the way it must have been a hundred or two hundred years ago. Hardly any people and hardly any loneliness. I'm undoubtedly overgeneralizing, but if the proper qualifications were introduced it would be true.

Technology is blamed for a lot of this loneliness, since the loneliness is certainly associated with the newer technological devices—TV, jets, freeways and so on—but I hope it's been made plain that the real evil isn't the objects of technology but the tendency of technology to isolate people into lonely attitudes of objectivity. It's the objectivity, the dualistic way of looking at things underlying technology, that produces the evil. That's why I went to so much trouble to show how technology could be used to destroy the evil. A person who knows how to fix motorcycles—with Quality—is less likely to run short of friends than one who doesn't. And they aren't going to see him as some kind of *object* either. Quality destroys objectivity every time.

Or if he takes whatever dull job he's stuck with—and they are all, sooner or later, dull—and, just to keep himself amused, starts to look for options of Quality, and secretly pursues these options, just for their own sake, thus making an art out of what he is doing, he's likely to discover that he becomes a much more interesting person and much less of an object to the people around him because his Quality decisions change *him* too. And not only the job and him, but others too because the Quality tends to fan out like waves. The Quality job he didn't think anyone was going to see *is* seen, and the person who sees it feels a little better because of it, and is likely to pass that feeling on to others, and in that way the Quality tends to keep on going.

My personal feeling is that this is how any further

improvement of the world will be done: by individuals making Quality decisions and that's all. God, I don't want to have any more enthusiasm for big programs full of social planning for big masses of people that leave individual Quality out. These can be left alone for a while. There's a place for them but they've got to be built on a foundation of Quality within the individuals involved. We've had that individual Quality in the past, exploited it as a natural resource without knowing it, and now it's just about depleted. Everyone's just about out of gumption. And I think it's about time to return to the rebuilding of *this* American resource—individual worth. There are political reactionaries who've been saying something close to this for years. I'm not one of them, but to the extent they're talking about real individual worth and not just an excuse for giving more money to the rich, they're right. We *do* need a return to individual integrity, self-reliance and old-fashioned gumption. We really do. I hope that in this Chautauqua some directions have been pointed to.

Phaedrus went a different path from the idea of individual, personal Quality decisions. I think it was a wrong one, but perhaps if I were in his circumstances I would go his way too. He felt that the solution started with a new philosophy, or he saw it as even broader than that—a new spiritual *rationality*—in which the ugliness and the loneliness and the spiritual blankness of dualistic technological reason would become illogical. Reason was no longer to be "value free." Reason was to be subordinate, logically, to Quality, and he was sure he would find the cause of its not being so back among the ancient Greeks, whose mythos had endowed our culture with the tendency underlying all the evil of our technology, the tendency *to do what is "reasonable" even when it isn't any good.* That was the root of the whole thing. Right there. I said a long time ago that he was in pursuit of the ghost of reason. This is what I meant. Reason and Quality had become separated and in conflict with each other and Quality had been forced under and reason made supreme somewhere back then.

It's begun to rain a little. Not so much we have to stop though. Just the faint beginnings of a drizzle.

The road leads out of the tall forests now and into open grey skies. Along the road are many billboards. Schenley's in warm-painted colors goes on forever, but one gets the feeling that Irma's gives tired, mediocre permanents because of the way the paint is cracking on her sign.

I have since read Aristotle again, looking for the massive evil that appears in the fragments from Phaedrus, but have not found it there. What I find in Aristotle is mainly a quite dull collection of generalizations, many of which seem impossible to justify in the light of modern knowledge, whose organization appears extremely poor, and which seems primitive in the way old Greek pottery in the museums seems primitive. I'm sure if I knew a lot more about it I would see a lot more and not find it primitive at all. But without knowing all that I can't see that it lives up either to the raves of the Great Books group or the rages of Phaedrus. I certainly don't see Aristotle's works as a major source of either positive or negative values. But the raves of the Great Books group are well known and published. Phaedrus' rages aren't, and it becomes part of my obligation to dwell on these.

Rhetoric is an art, Aristotle began, *because it can be reduced to a rational system of order.*

That just left Phaedrus aghast. Stopped. He'd been prepared to decode messages of great subtlety, systems of great complexity in order to understand the deeper inner meaning of Aristotle, claimed by many to be the greatest philosopher of all time. And then to get hit, right off, straight in the face, with an asshole statement like that! It really shook him.

He read on:

Rhetoric can be subdivided into particular proofs and topics on the one hand and common proofs on the other. The particular proofs can be subdivided into methods of proof and kinds of proof. The methods of proofs are the artificial proofs and the inartificial proofs. Of the artificial proofs there are ethical proofs, emotional proofs and logical proofs. Of the ethical proofs there are practical wisdom, virtue and good will. The particular methods employing artificial proofs of the ethical kind involving good will require a knowledge of the emotions, and for those who have forgotten what these are, Aristotle provides a list. They are anger, slight (subdivisible into contempt, spite and insolence), mildness, love or friendship, fear, confidence, shame, shamelessness, favor, benevolence, pity, virtuous indignation, envy, emulation and contempt.

Remember the description of the motorcycle given way back in South Dakota? The one which carefully enumerated all the motorcycle parts and functions? Recognize the similarity? Here, Phaedrus was convinced, was the originator of that style of discourse.

For page after page Aristotle went on like this. Like some third-rate technical instructor, naming everything, showing the relationships among the things named, cleverly inventing an occasional new relationship among the things named, and then waiting for the bell so he can get on to repeat the lecture for the next class.

Between the lines Phaedrus read no doubts, no sense of awe, only the eternal smugness of the professional academician. Did Aristotle really think his students would be better rhetoricians for having learned all these endless names and relationships? And if not, did he really think he was teaching rhetoric? Phaedrus thought that he really did. There was nothing in his style to indicate that Aristotle was ever one to doubt Aristotle. Phaedrus saw Aristotle as tremendously satisfied with this neat little stunt of naming and classifying everything. His world began and ended with this stunt. The reason why, if he were not more than two thousand years dead, he would have gladly rubbed him out is that he saw him as a prototype for the many millions of self-satisfied and truly ignorant teachers throughout history who have smugly and callously killed the creative spirit of their students with this dumb ritual of analysis, this blind, rote, eternal naming of things. Walk into any of a hundred thousand classrooms today and hear the teachers divide and subdivide and interrelate and establish "principles" and study "methods" and what you will hear is the ghost of Aristotle speaking down through the centuries—the desiccating lifeless voice of dualistic reason.

The sessions on Aristotle were round an enormous wooden round table in a dreary room across the street from a hospital, where the late-afternoon sun from over the hospital roof hardly penetrated the window dirt and polluted city air beyond. Wan and pale and depressing. During the middle of the hour he noticed that this enormous table had a huge crack that ran right across it near the middle. It looked as though it had been there for years, but that no one had thought to repair it. Too busy, no doubt, with more important things. At the end of the hour he finally asked, "May questions about Aristotle's rhetoric be asked?"

"If you have read the material," he was told. He noticed in the eye of the Professor of Philosophy the same set he had seen the first day of registration. He took warning from it that he had better read the material very thoroughly, and did so.

The rain comes down more heavily now and we stop to snap on the face mask to the helmet. Then we go again at moderate speed. I watch for chuckholes, sand and grease slicks.

The next week Phaedrus had read the material and was prepared to take apart the statement that rhetoric is an art because it can be reduced to a rational system of order. By this criterion General Motors produced pure art, whereas Picasso did not. If there were deeper meanings to Aristotle than met the eye this would be as good a place as any to make them visible.

But the question never got raised. Phaedrus put up his hand to do so, caught a microsecond flash of malice from the teacher's eye, but then another student said, almost as an interruption, "I think there are some very dubious statements here."

That was all he got out.

"Sir, we are not here to *learn* what *you* think!" hissed the Professor of Philosophy. Like acid. "We are here to learn what *Aristotle* thinks!" Straight in the face. "When we wish to learn what you think we will assign a course in the subject!"

Silence. The student is stunned. So is everyone else.

But the Professor of Philosophy is not done. He points his finger at the student and demands, "According to Aristotle: What are the three kinds of particular rhetoric according to subject matter discussed?"

More silence. The student doesn't know. "Then you haven't *read* it, have you?"

And now, with a gleam that indicates he has intended this all along, the Professor of Philosophy swings his finger around and points it at Phaedrus.

"You, sir, what are the three kinds of particular rhetoric according to subject matter discussed?"

But Phaedrus is prepared. "Forensic, deliberative and epideictic," he answers calmly.

"What are the epideictic techniques?"

"The technique of identifying likenesses, the technique of praise, that of encomium and that of amplification."

"Yaaas . . ." says the Professor of Philosophy slowly. Then all is silent.

The other students looked shocked. They wonder what has happened. Only Phaedrus knows, and perhaps the Professor of Philosophy. An innocent student has caught blows intended for him.

Now everyone's face becomes carefully composed in defense against more of this sort of questioning. The Professor of Philosophy has made a mistake. He's wasted his disciplinary authority on an innocent student while Phaedrus, the guilty one, the hostile one, is still at large. And getting larger and larger. Since he has asked no questions there is now no way to cut him down. And now that he sees how the questions will be answered he's certainly not about to ask them.

The innocent student stares down at the table, face red, hands shrouding his eyes. His shame becomes Phaedrus' anger. In all his classes he never once talked to a student like that. So that's how they teach classics at the University of Chicago. Phaedrus knows the Professor of Philosophy now. But the Professor of Philosophy doesn't know Phaedrus.

The grey rainy skies and sign-strewn road descend to Crescent City, California, grey and cold and wet, and Chris and I look and see the water, the ocean, in the distance beyond piers and grey buildings. I remember this was our great goal all these days. We enter a restaurant with a fancy red carpet and fancy menus with extremely high prices. We are the only people here. We eat silently, pay and are on the road again, south now, cold and misty.

In the next sessions the shamed student is no longer present. No surprise. The class is completely frozen, as is inevitable when an incident like that has taken place. Each session, just one person does all the talking, the Professor of Philosophy, and he talks and talks and talks to faces that have turned into masks of neutrality.

The Professor of Philosophy seems quite aware of what has happened. His previous little eye-flick of malice toward Phaedrus has turned to a little eye-flick of fear. He seems to understand that within the present classroom situation, when the time comes, he can get exactly the same treatment he gave, and there will be no sympathy from any of the faces before him. He's thrown away his right to courtesy. There's no way to prevent retaliation now except to keep covered.

But to keep covered he must work hard, and say things exactly right. Phaedrus understands this too. By remaining silent he can now learn under what are very advantageous circumstances.

Phaedrus studied hard during this period, and learned extremely fast, and kept his mouth shut, but it would be wrong to give the least impression that he was any sort of good student. A good student seeks knowledge fairly and impartially. Phaedrus did not. He had an axe to grind and all he sought were those things that helped him grind it, and the means of knocking down anything which prevented him from grinding it. He had no time for or interest in other people's Great Books. He was there solely to write a Great Book of his own. His attitude toward Aristotle was grossly unfair for the same reason Aristotle was unfair to *his* predecessors. They fouled up what he wanted to say.

Aristotle fouled up what Phaedrus wanted to say by placing rhetoric in an outrageously minor category in his hierarchic order of things. It was a branch of Practical Science, a kind of shirttail relation to the *other* category, Theoretical Science, which Aristotle was mainly involved in. As a branch of Practical Science it was isolated from any concern with Truth or Good or Beauty, except as devices to throw into an argument. Thus Quality, in Aristotle's system, is totally divorced from rhetoric. This contempt for rhetoric, combined with Aristotle's *own* atrocious quality of rhetoric, so completely alienated Phaedrus he couldn't read anything Aristotle said without seeking ways to despise it and attack it.

This was no problem. Aristotle has always been eminently attackable and eminently attacked throughout history, and shooting down Aristotle's patent absurdities, like shooting fish in a barrel, didn't afford much satisfaction. If he hadn't been so partial Phaedrus might have learned some valuable Aristotelian techniques of bootstrapping oneself into new areas of knowledge, which was what the committee was really set up for. But if he hadn't been so partial in his search for a place to launch his work on Quality, he wouldn't have been there in the first place, so it really didn't have any chance to work out at all.

The Professor of Philosophy lectured, and Phaedrus listened to both the classic form and romantic surface of what was said. The Professor of Philosophy seemed most ill at ease on the subject of "dialectic." Although Phaedrus couldn't figure out why in terms of classic form, his growing romantic sensitivity told him he was on the scent of something—a quarry.

Dialectic, eh?

Aristotle's book had begun with it, in a most mystifying way. Rhetoric is a counterpart of dialectic, it had said, as if this were of the greatest importance, yet why this was so important was never explained. It was followed with a number of other disjointed statements, which gave the impression that a great deal had been left out, or the material had been assembled wrongly, or the printer had left something out, because no matter how many times he read it nothing jelled. The only thing that was clear was that Aristotle was very much concerned about the relation of rhetoric to dialectic. To Phaedrus' ear, the same ill ease he had observed in the Professor of Philosophy appeared here.

The Professor of Philosophy had defined dialectic, and Phaedrus had listened carefully, but it was in one ear and out the other, a characteristic that philosophic statements often have when something is left out. In a later class another student who seemed to be having the same trouble asked the Professor of Philosophy to redefine dialectic and this time the Professor had glanced at Phaedrus with another quick flicker of fear and become *very* edgy. Phaedrus began to wonder if "dialectic" had some special meaning that made it a fulcrum word—one that can shift the balance of an argument, depending on how it's placed. It was.

Dialectic generally means "of the nature of the dialogue," which is a conversation between two persons. Nowadays it means logical argumentation. It involves a technique of cross-examination, by which truth is arrived at. It's the mode of discourse of Socrates in the *Dialogues* of Plato. Plato believed the dialectic was the sole method by which the truth was arrived at. The only one.

That's why it's a fulcrum word. Aristotle attacked this belief, saying that the dialectic was only suitable for some purposes—to inquire into men's beliefs, to arrive at truths about eternal forms of things, known as *Ideas,* which were fixed and unchanging and constituted reality for Plato. Aristotle said there is also the method of science, or "physical" method, which observes physical facts and arrives at truths about substances, which undergo change. This duality of form and substance and the scientific method of arriving at facts about substances were central to Aristotle's philosophy. Thus the dethronement of dialectic from what Socrates and Plato held it to be was absolutely essential for Aristotle, and "dialectic" was and still is a fulcrum word.

Phaedrus guessed that Aristotle's diminution of dialectic, from Plato's sole method of arriving at truth to a "counterpart of rhetoric," might be as infuriating to modern Platonists as it would have been to Plato. Since the Professor of Philosophy didn't know what Phaedrus' "position" was, this was what was making him edgy. He might be afraid that Phaedrus the Platonist was going to jump him. If so, he certainly had nothing to worry about. Phaedrus wasn't insulted that dialectic had been brought down to the level of rhetoric. He was outraged that rhetoric had been brought down to the level of dialectic. Such was the confusion at the time.

The person to clear all this up, of course, was Plato, and fortunately he was the next to appear at the round table with the crack running across the middle in the dim dreary room across from the hospital building in South Chicago.

We follow the coast now, cold, wet and depressed. The rain has let up, temporarily, but the sky shows no hope. At one point I see a beach and some people walking on it in the wet sand. I'm tired and so I stop.

As he gets off, Chris says, "What are we stopping for?"

"I'm tired," I say. The wind blows cold off the ocean and where it has formed dunes, now wet and dark from the rain that must just have ended here, I find a place to lie down, and this makes me a little warmer.

I don't sleep though. A little girl appears over the top of the dune looking as though she wants me to come and play. After a while she goes away.

In time Chris comes back and wants to go. He says he has found some funny plants out on the rocks that have feelers which pull in when you touch them. I go with him and see between rises of waves on the rocks that they are sea anemones, which are not plants but animals. I tell him the tentacles can paralyze small fish. The tide must be all the way out or we wouldn't see these, I say. From the corner of my eye I see the little girl on the other side of the rocks has picked up a starfish. Her parents are carrying some starfish too.

We get on the motorcycle and move south. Sometimes the rain gets heavy and I snap on the bubble so it doesn't sting my face, but I don't like this and take it off when the rain dies away. We should reach Arcata before dark but I don't want to go too fast on this wet road.

I think it was Coleridge who said everyone is either a Platonist or an Aristotelian. People who can't stand Aristotle's endless specificity of detail are natural lovers of Plato's soaring generalities. People who can't stand the eternal lofty idealism of Plato welcome the down-to-earth facts of Aristotle. Plato is the essential Buddha-seeker who appears again and again in each generation, moving onward and upward toward the "one." Aristotle is the eternal motorcycle mechanic who prefers the "many." I myself am pretty much Aristotelian in this sense, preferring to find the Buddha in the quality of the facts around me, but Phaedrus was clearly a Platonist by temperament and when the classes shifted to Plato he was greatly relieved. His Quality and Plato's Good were so similar that if it hadn't been for some notes Phaedrus left I might have thought they were identical. But he denied it, and in time I came to see how important this denial was.

The course in the Analysis of Ideas and Study of Methods was not concerned with Plato's notion of the Good, however; it was concerned with Plato's notion of rhetoric. Rhetoric, Plato spells out very clearly, is in no way connected with the Good; rhetoric is "the Bad." The people Plato hates most, next to tyrants, are rhetoricians.

The first of the Platonic *Dialogues* assigned is the *Gorgias,* and Phaedrus has a sense of having arrived. This at last is where he wants to be.

All along he has had a feeling of being swept forward by forces he doesn't understand—Messianic forces. October has come and gone. Days have become phantasmal and incoherent, except in terms of Quality. Nothing matters except that he has a new and shattering and world-shaking truth about to be born, and like it or not, the world is morally obligated to accept it.

In the dialogue, Gorgias is the name of a Sophist whom Socrates cross-examines. Socrates knows very well what Gorgias does for a living and how he does it, but he starts his Twenty Questions dialectic by asking Gorgias with what rhetoric is concerned. Gorgias answers that it is concerned with discourse. In answer to another question he says that its end is to persuade. In answer to another question he says its place is in the law courts and other assemblies. And in answer to still another question he says its subject is the just and the unjust. All this, which is simply Gorgias' description of what people called Sophists have tended to do, now becomes subtly rendered by Socrates' dialectic into something else. Rhetoric has become an object, and as an object has parts. And the parts have relationships to one another and these relations are immutable. One sees quite clearly in this dialogue how the analytic knife of Socrates hacks Gorgias' art into pieces. What is even more important, one sees that the pieces are the basis of Aristotle's art of rhetoric.

Socrates had been one of Phaedrus' childhood heroes and it shocked and angered him to see this dialogue taking place. He filled the margins of the text with answers of his own. These must have frustrated him greatly, because there was no way of knowing how the dialogue would have gone if these answers had been made. At one place Socrates asks to what class of things do the words which Rhetoric uses relate. Gorgias answers, "The Greatest and the Best." Phaedrus, no doubt recognizing Quality in this answer, has written "True!" in the margin. But Socrates responds that this answer is ambiguous. He is still in the dark. "Liar!" writes Phaedrus in the margin, and he cross-references a page in another dialogue where Socrates makes it clear he could *not* have been "in the dark."

Socrates is not using dialectic to understand rhetoric, he is using it to destroy it, or at least to bring it into disrepute, and so his questions are not real questions at all—they are just word-traps which Gorgias and his fellow rhetoricians fall into. Phaedrus is quite incensed by all this and wishes he were there.

In class, the Professor of Philosophy, noting Phaedrus' apparent good behavior and diligence, has decided he may not be such a bad student after all. This is a second mistake. He has decided to play a little game with Phaedrus by asking him what he thinks of cookery. Socrates has demonstrated to Gorgias that both rhetoric and cooking are branches of pandering— pimping—because they appeal to the emotions rather than true knowledge.

In response to the Professor's question, Phaedrus gives Socrates' answer that cookery is a branch of pandering.

There's a titter from one of the women in the class which displeases Phaedrus because he knows the Professor is trying for a dialectical hold on him similar to the kind Socrates gets on his opponents, and his answer is not intended to be funny but simply to throw off the dialectical hold the Professor is trying to get. Phaedrus is quite ready to recite in detail the exact arguments Socrates uses to establish this view.

But that isn't what the Professor wants. He wants to have a dialectical discussion in class in which he, Phaedrus, is the rhetorician and is thrown by the force of dialectic. The Professor frowns and tries again. "No. I mean, do you really think that a well-cooked meal served in the best of restaurants is really something that we should turn down?"

Phaedrus asks, "You mean my *personal* opinion?" For months now, since the innocent student disappeared, there have been no personal opinions ventured in this class.

"Yaaas," the Professor says.

Phaedrus is silent and tries to work out an answer. Everyone is waiting. His thoughts move up to lightning speed, winnowing through the dialectic, playing one argumentative chess opening after another, seeing that each one loses, and moving to the next one, faster and faster—but all the class witnesses is silence. Finally, in embarrassment, the Professor drops the question and begins the lecture.

But Phaedrus doesn't hear the lecture. His mind races on and on, through the permutations of the dialectic, on and on, hitting things, finding new branches and sub-branches, exploding with anger at each new discovery of the viciousness and meanness and lowness of this "art" called dialectic. The Professor, looking at his expression, becomes quite alarmed, and continues the lecture in a kind of panic. Phaedrus' mind races on and on and then on further, seeing now at last a kind of evil thing, an evil deeply entrenched in himself, which *pretends* to try to understand love and beauty and truth and wisdom but whose real purpose is never to understand them, whose real purpose is always to usurp them and enthrone itself. Dialectic—the usurper. That is what he sees. The parvenu, muscling in on all that is Good and seeking to contain it and control it. Evil. The Professor calls the lecture to an early end and leaves the room hurriedly.

After the students have filed out silently Phaedrus sits alone at the huge round table until the sun through the sooty air beyond the window disappears and the room becomes grey and then dark.

The next day he is at the library waiting for it to open and when it does he begins to read furiously, back *behind* Plato for the first time, into what little is known of those rhetoricians he so despised. And what he discovers begins to confirm what he has already intuited from his thoughts the evening before.

Plato's condemnation of the Sophists is one which many scholars have already taken with great misgivings. The Chairman of the committee himself has suggested that critics who are not certain what Plato meant should be equally uncertain of what Socrates' antagonists in the dialogues meant. When it is known that Plato put his own words in Socrates' mouth (Aristotle says this) there should be no reason to doubt that he could have put his own words into other mouths too.

Fragments by other ancients seemed to lead to other evaluations of the Sophists. Many of the older Sophists were selected as "ambassadors" of their cities, certainly no office of disrespect. The name Sophist was even applied without disparagement to Socrates and Plato themselves. It has even been suggested by some later historians that the reason Plato hated the Sophists so was that they could not compare with his master, Socrates, who was in actuality the greatest Sophist of them all. This last explanation is interesting, Phaedrus thinks, but unsatisfactory. You don't abhor a school of which your master is a member. What was Plato's *real* purpose in this? Phaedrus reads further and further into pre-Socratic Greek thought to find out, and eventually comes to the view that Plato's hatred of the rhetoricians was part of a much larger struggle in which the reality of the Good, represented by the Sophists, and the reality of the True, represented by the dialecticians, were engaged in a huge struggle for the future mind of man. Truth won, the Good lost, and that is why today we have so little difficulty accepting the reality of truth and so much difficulty accepting the reality of Quality, even though there is no more agreement in one area than in the other.

To understand how Phaedrus arrives at this requires some explanation:

One must first get over the idea that the time span between the last caveman and the first Greek philosophers was short. The absence of any history for this period sometimes gives this illusion. But before the Greek philosophers arrived on the scene, for a period of at least *five times* all our recorded history since the Greek philosophers, there existed civilizations in an advanced state of development. They had villages and cities, vehicles, houses, marketplaces, bounded fields, agricultural implements and domestic animals, and led a life quite as rich and varied as that in most rural areas of the world today. And like people in those areas today they saw no reason to write it all down, or if they did, they wrote it on materials that have never been found. Thus we know nothing about them. The "Dark Ages" were merely the resumption of a natural way of life that had been momentarily interrupted by the Greeks.

Early Greek philosophy represented the first conscious search for what was imperishable in the affairs of men. Up to then what was imperishable was within the domain of the Gods, the myths. But now, as a result of the growing impartiality of the Greeks to the world around them, there was an increasing power of abstraction which permitted them to regard the old Greek mythos not as revealed truth but as imaginative creations of art. This consciousness, which had never existed anywhere before in the world, spelled a whole new level of transcendence for the Greek civilization.

But the mythos goes on, and that which destroys the old mythos becomes the new mythos, and the new mythos under the first Ionian philosophers became transmuted into philosophy, which enshrined permanence in a new way. Permanence was no longer the exclusive domain of the Immortal Gods. It was also to be found within Immortal Principles, of which our current law of gravity has become one.

The Immortal Principle was first called water by Thales. Anaximenes called it air. The Pythagoreans called it number and were thus the first to see the Immortal Principle as something nonmaterial. Heraclitus called the Immortal Principle fire and introduced change as part of the Principle. He said the world exists as a conflict and tension of opposites. He said there is a One and there is a Many and the One is the universal law which is immanent in all things. Anaxagoras was the first to identify the One as *nous,* meaning "mind."

Parmenides made it clear for the first time that the Immortal Principle, the One, Truth, God, is separate from appearance and from opinion, and the importance of this separation and its effect upon subsequent history cannot be overstated. It's here that the classic mind, for the first time, took leave of its romantic origins and said, "The Good and the True are not necessarily the same," and goes its separate way. Anaxagoras and Parmenides had a listener named Socrates who carried their ideas into full fruition.

What is essential to understand at this point is that until now there was no such thing as mind and matter, subject and object, form and substance. Those divisions are just dialectical inventions that came later. The modern mind sometimes tends to balk at the thought of these dichotomies being inventions and says, "Well, the divisions were *there* for the Greeks to discover," and you have to say, "*Where* were they? Point to them!" And the modern mind gets a little confused and wonders what this is all about anyway, and *still* believes the divisions were there.

But they weren't, as Phaedrus said. They are just ghosts, immortal gods of the modern mythos which appear to us to be real because we are *in* that mythos. But in reality they are just as much an artistic creation as the anthropomorphic Gods they replaced.

The pre-Socratic philosophers mentioned so far all sought to establish a universal Immortal Principle in the external world they found around them. Their common effort united them into a group that may be called Cosmologists. They all agreed that such a principle existed but their disagreements as to what it was seemed irresolvable. The followers of Heraclitus insisted the Immortal Principle was change and motion. But Parmenides' disciple, Zeno, proved through a series of paradoxes that any perception of motion and change is illusory. Reality had to be motionless.

The resolution of the arguments of the Cosmologists came from a new direction entirely, from a group Phaedrus seemed to feel were early humanists. They were teachers, but what they sought to teach was not principles, but beliefs of men. Their object was not any single absolute truth, but the improvement of men. All principles, all truths, are relative, they said. "Man is the measure of all things." These were the famous teachers of "wisdom," the Sophists of ancient Greece.

To Phaedrus, this backlight from the conflict between the Sophists and the Cosmologists adds an entirely new dimension to the *Dialogues* of Plato. Socrates is not just expounding noble ideas in a vacuum. He is in the middle of a war between those who think truth is absolute and those who think truth is relative. He is fighting that war with everything he has. The Sophists are the enemy.

Now Plato's hatred of the Sophists makes sense. He and Socrates are defending the Immortal Principle of the Cosmologists against what they consider to be the decadence of the Sophists. Truth. Knowledge. That which is independent of what anyone thinks about it. The ideal that Socrates died for. The ideal that Greece alone possesses for the first time in the history of the world. It is still a very fragile thing. It can disappear completely. Plato abhors and damns the Sophists without restraint, not because they are low and immoral people—there are obviously much lower and more immoral people in Greece he completely ignores. He damns them because they threaten mankind's first beginning grasp of the idea of truth. That's what it is all about.

The results of Socrates' martyrdom and Plato's unexcelled prose that followed are nothing less than the whole world of Western man as we know it. If the idea of truth had been allowed to perish unrediscovered by the Renaissance it's unlikely that we would be much beyond the level of prehistoric man today. The ideas of science and technology and other systematically organized efforts of man are dead-centered on it. It is the nucleus of it all.

And yet, Phaedrus understands, what he is saying about Quality is somehow opposed to all this. It seems to agree much more closely with the Sophists.

"Man is the measure of all things." Yes, that's what he is saying about Quality. Man is not the *source* of all things, as the subjective idealists would say. Nor is he

the passive observer of all things, as the objective idealists and materialists would say. The Quality which
creates the world emerges as a *relationship* between
man and his experience. He is a *participant* in the creation of all things. The *measure* of all things—it fits.
And they taught rhetoric—that fits.

The one thing that doesn't fit what he says and what
Plato said about the Sophists is their profession of
teaching *virtue.* All accounts indicate this was absolutely central to their teaching, but how are you
going to teach virtue if you teach the relativity of all
ethical ideas? Virtue, if it implies anything at all, implies an ethical absolute. A person whose idea of what
is proper varies from day to day can be admired for his
broadmindedness, but not for his *virtue.* Not, at least,
as Phaedrus understands the word. And how could they
get *virtue* out of rhetoric? This is never explained anywhere. Something is missing.

His search for it takes him through a number of his-
tories of ancient Greece, which as usual he reads detective style, looking only for facts that may help him
and discarding all those that don't fit. And he is reading H. D. F. Kitto's *The Greeks,* a blue and white paperback which he has bought for fifty cents, and he has
reached a passage that describes "the very soul of the
Homeric hero," the legendary figure of predecadent,
pre-Socratic Greece. The flash of illumination that follows these pages is so intense the heroes are never
erased and I can see them with little effort of recall.

The *Iliad* is the story of the siege of Troy, which will
fall in the dust, and of its defenders who will be killed
in battle. The wife of Hector, the leader, says to him:
"Your strength will be your destruction; and you have

no pity either for your infant son or for your unhappy wife who will soon be your widow. For soon the Acheans will set upon you and kill you; and if I lose you it would be better for me to die."

Her husband replies:

"Well do I know this, and I am sure of it: that day is coming when the holy city of Troy will perish, and Priam and the people of wealthy Priam. But my grief is not so much for the Trojans, nor for Hecuba herself, nor for Priam the King, nor for my many noble brothers, who will be slain by the foe and will lie in the dust, as for you, when one of the bronze-clad Acheans will carry you away in tears and end your days of freedom. Then you may live in Argos, and work at the loom in another woman's house, or perhaps carry water for a woman of Messene or Hyperia, sore against your will: but hard compulsion will lie upon you. And then a man will say as he sees you weeping. 'This was the wife of Hector, who was the noblest in battle of the horse-taming Trojans, when they were fighting around Ilion.' This is what they will say: and it will be fresh grief for you, to fight against slavery bereft of a husband like that. But may I be dead, may the earth be heaped over my grave before I hear your cries, and of the violence done to you."

So spake shining Hector and held out his arms to his son. But the child screamed and shrank back into the bosom of the well-girdled nurse, for he took fright at the sight of his dear father—at the bronze and the crest of the horsehair which he saw swaying terribly from the top of the helmet. His father laughed aloud, and his lady mother too. At once shining Hector took the helmet off his head and laid it on the ground, and when he had kissed his

> dear son and dandled him in his arms, he prayed to Zeus
> and to the other Gods: Zeus and ye other Gods, grant that
> this my son may be, as I am, most glorious among the
> Trojans and a man of might, and greatly rule in Ilion.
> And may they say, as he returns from war, 'He is far bet-
> ter than his father.'

"What moves the Greek warrior to deeds of hero-
ism," Kitto comments, "is not a sense of duty as we understand it—duty towards others: it is rather duty
towards himself. He strives after that which we translate 'virtue' but is in Greek *aretê,* 'excellence' . . . we
shall have much to say about *aretê.* It runs through
Greek life."

There, Phaedrus thinks, is a definition of Quality
that had existed a thousand years before the dialecticians ever thought to put it to word-traps. Anyone who
cannot understand this meaning without logical
definiens and *definendum* and *differentia* is either lying
or so out of touch with the common lot of humanity as
to be unworthy of receiving any reply whatsoever.
Phaedrus is fascinated too by the description of the
motive of "duty toward self" which is an almost exact
translation of the Sanskrit word *dharma,* sometimes
described as the "one" of the Hindus. Can the *dharma*
of the Hindus and the "virtue" of the ancient Greeks be
identical?

Then Phaedrus feels a tugging to read the passage
again, and he does so and then . . . what's this?! . . .
*"That which we translate 'virtue' but is in Greek 'ex-
cellence.' "*

Lightning hits!

Quality! Virtue! Dharma! That is what the Sophists

were teaching! *Not* ethical relativism. *Not* pristine "virtue." But *aretê*. Excellence. *Dharma!* Before the Church of Reason. Before substance. Before form. Before mind and matter. Before dialectic itself. Quality had been absolute. Those first teachers of the Western world were teaching *Quality,* and the medium they had chosen was that of rhetoric. He has been doing it right all along.

The rain has lifted enough so that we can see the horizon now, a sharp line demarking the light grey of the sky and the darker grey of the water.

Kitto had more to say about this *aretê* of the ancient Greeks. "When we meet *aretê* in Plato," he said, "we translate it 'virtue' and consequently miss all the flavour of it. 'Virtue,' at least in modern English, is almost entirely a moral word; *aretê,* on the other hand, is used indifferently in all the categories, and simply means excellence."

> Thus the hero of the *Odyssey* is a great fighter, a wily schemer, a ready speaker, a man of stout heart and broad wisdom who knows that he must endure without too much complaining what the gods send; and he can both build and sail a boat, drive a furrow as straight as anyone, beat a young braggart at throwing the discus, challenge the Pheacian youth at boxing, wrestling or running; flay, skin, cut up and cook an ox, and be moved to tears by a song. He is in fact an excellent all-rounder; he has surpassing *aretê.*
>
> *Aretê* implies a respect for the wholeness or oneness of life, and a consequent dislike of specialization. It im-

plies a contempt for efficiency—or rather a much higher idea of efficiency, an efficiency which exists not in one department of life but in life itself.

Phaedrus remembered a line from Thoreau: "You never gain something but that you lose something." And now he began to see for the first time the unbelievable magnitude of what man, when he gained power to understand and rule the world in terms of dialectic truths, had lost. He had built empires of scientific capability to manipulate the phenomena of nature into enormous manifestations of his own dreams of power and wealth—but for this he had exchanged an empire of understanding of equal magnitude: an understanding of what it is to be a part of the world, and not an enemy of it.

One can acquire some peace of mind from just watching that horizon. It's a geometer's line . . . completely flat, steady and known. Perhaps it's the original line that gave rise to Euclid's understanding of lineness; a reference line from which was derived the original calculations of the first astronomers that charted the stars.

Phaedrus knew, with the same mathematical assurance Poincaré had felt when he resolved the Fuchsian equations, that this Greek *aretê* was the missing piece that completed the pattern, but he read on now for completion.

The halo around the heads of Plato and Socrates is now gone. He sees that they consistently are doing exactly that which they accuse the Sophists of doing—using emotionally persuasive language for the ulterior purpose of making the weaker argument, the case for dialectic, appear the stronger. We always condemn most in others, he thought, that which we most fear in ourselves.

But why? Phaedrus wondered. Why destroy *aretê*? And no sooner had he asked the question than the answer came to him. Plato *hadn't* tried to destroy *aretê*. He had *encapsulated* it; made a permanent, fixed Idea out of it; had *converted* it to a rigid, immobile Immortal Truth. He made *aretê* the Good, the highest form, the highest Idea of all. It was subordinate only to Truth itself, in a synthesis of all that had gone before.

That was why the Quality that Phaedrus had arrived at in the classroom had seemed so close to Plato's Good. Plato's Good was *taken* from the rhetoricians. Phaedrus searched, but could find no previous cosmologists who had talked about the Good. That was from the Sophists. The difference was that Plato's Good was a fixed and eternal and unmoving Idea, whereas for the rhetoricians it was not an Idea at all. The Good was not a *form* of reality. It was reality itself, ever changing, ultimately unknowable in any kind of fixed, rigid way.

Why had Plato done this? Phaedrus saw Plato's philosophy as a result of *two* syntheses.

The first synthesis tried to resolve differences between the Heraclitans and the followers of Parmenides. Both Cosmological schools upheld Immortal Truth. In order to win the battle for Truth in which *aretê* is subordinate, against his enemies who would teach *aretê* in which truth is subordinate, Plato must first resolve the internal conflict among the Truth-believers. To do this he says that Immortal Truth is not just change, as the followers of Heraclitus said. It is not just changeless being, as the followers of Parmenides said. Both these

Immortal Truths coexist as Ideas, which are change-less, and Appearance, which changes. This is why Plato finds it necessary to separate, for example, "horse-ness" from "horse" and say that horseness is real and fixed and true and unmoving, while the horse is a mere, unimportant, *transitory* phenomenon. Horseness is pure Idea. The horse that one sees is a collection of changing Appearances, a horse that can flux and move around all it wants to and even die on the spot without disturbing horseness, which is the Immortal Principle and can go on forever in the path of the Gods of old.

Plato's second synthesis is the incorporation of the Sophists' *aretê* into this dichotomy of Ideas and Appearance. He gives it the position of highest honor, subordinate only to Truth itself and the method by which Truth is arrived at, the dialectic. But in his attempt to unite the Good and the True by making the Good the highest Idea of all, Plato is nevertheless usurping *aretê*'s place with dialectically determined truth. Once the Good has been contained as a dialectical idea it is no trouble for another philosopher to come along and show by dialectical methods that *aretê,* the Good, can be more advantageously demoted to a lower position within a "true" order of things, more compatible with the inner workings of dialectic. Such a philosopher was not long in coming. His name was Aristotle.

Aristotle felt that the mortal horse of Appearance which ate grass and took people places and gave birth to little horses deserved far more attention than Plato was giving it. He said that the horse is not mere Appearance. The Appearances cling to something which is independent of them and which, like Ideas, is unchanging. The "something" that Appearances cling to he named "substance." And at that moment, and not until that moment, our modern scientific understanding of reality was born.

Under Aristotle the "Reader," whose knowledge of Trojan *aretê* seems conspicuously absent, forms and substances dominate all. The Good is a relatively minor branch of knowledge called ethics; reason, logic, knowledge are his primary concerns. *Aretê* is dead and science, logic and the University as we know it today have been given their founding charter: to find and invent an endless proliferation of forms about the substantive elements of the world and call these forms knowledge, and transmit these forms to future generations. As "the system."

And rhetoric. Poor rhetoric, once "learning" itself, now becomes reduced to the teaching of mannerisms and forms, Aristotelian forms, for writing, as if these mattered. Five spelling errors, Phaedrus remembered, *or* one error of sentence completeness, *or* three misplaced modifiers, *or* . . . it went on and on. Any of these was sufficient to inform a student that he did not know rhetoric. After all, that's what rhetoric is, isn't it? Of course there's "empty rhetoric," that is, rhetoric that has emotional appeal without proper subservience to dialectical truth, but we don't want any of *that,* do we? That would make us like those liars and cheats and defilers of ancient Greece, the Sophists—remember *them*? We'll learn the Truth in our other academic courses, and then learn a little rhetoric so that we can write it nicely and impress our bosses who will advance us to higher positions.

Forms and mannerisms—hated by the best, loved by

the worst. Year after year, decade after decade of little front-row "readers," mimics with pretty smiles and neat pens, out to get their Aristotelian A's while those who possess the real *aretê* sit silently in back of them wondering what is wrong with themselves that they cannot like this subject.

And today in those few Universities that bother to teach classic ethics anymore, students, following the lead of Aristotle and Plato, endlessly play around with the question that in ancient Greece never needed to be asked: "What is the Good? And how do we define it? Since different people have defined it differently, how can we *know* there is any good? Some say the good is found in happiness, but how do we know what happiness is? And how can happiness be defined? Happiness and good are not objective terms. We cannot deal with them scientifically. And since they aren't objective they just exist in your mind. So if you want to be happy just change your mind. Ha-ha, ha-ha."

Aristotelian ethics, Aristotelian definitions, Aristotelian logic, Aristotelian forms, Aristotelian substances, Aristotelian rhetoric, Aristotelian laughter . . . ha-ha, ha-ha.

And the bones of the Sophists long ago turned to dust and what they said turned to dust with them and the dust was buried under the rubble of declining Athens through its fall and Macedonia through its decline and fall. Through the decline and death of ancient Rome and Byzantium and the Ottoman Empire and the modern states—buried so deep and with such ceremoniousness and such unction and such evil that only a madman centuries later could discover the clues needed to uncover them, and see with horror what had been done. . . .

The road has become so dark I have to turn on my headlight now to follow it through these mists and rain.

30

At Arcata we enter a small diner, cold and wet, and eat chili and beans and drink coffee.

Then we are back on the road again, freeway now, fast and wet. We'll go to within an easy day's distance from San Francisco and then stop.

The freeway picks up strange reflections in the rain from oncoming lights across the median. The rain hits like pellets against the bubble, which refracts the lights in strange circular and then semicircular waves as they go by. Twentieth century. It's all around us now, this twentieth century. Time to finish this twentieth-century odyssey of Phaedrus and be done with it.

The next time the class in Ideas and Methods 251, Rhetoric, met at the large round table in South Chicago, a department secretary announced that the Professor of Philosophy was ill. The following week he was still ill. The somewhat bewildered remnants of the class, which had dwindled to a third of its size, went on their own across the street for coffee.

At the coffee table a student whom Phaedrus had marked as bright but intellectually snobbish said, "I consider this one of the most unpleasant classes I have ever been in." He seemed to look down on Phaedrus with womanish peevishness as a spoiler of what should have been a nice experience.

"I thoroughly agree," Phaedrus said. He waited for some sort of attack, but it didn't come.

The other students seemed to sense that Phaedrus was the cause of all this but they had nothing to go on. Then an older woman at the other end of the coffee table asked why he was attending the class.

"I'm in the process of trying to discover that," Phaedrus said.

"Do you attend full-time?" she asked.

"No, I teach full-time at Navy Pier."

"What do you teach?"

"Rhetoric."

She stopped talking and everyone at the table looked at him and became silent.

November wore on. The leaves, which had turned a beautiful sunlit orange in October, fell from the trees, leaving barren branches to meet the cold winds from the north. A first snow fell, then melted, and a drab city waited for winter to come.

In the Professor of Philosophy's absence, another Platonic dialogue had been assigned. Its title was *Phaedrus,* which meant nothing to *our* Phaedrus since he didn't call himself by that name. The Greek Phaedrus is not a Sophist but a young orator who is a foil for Socrates in this dialogue, which is about the nature of love and the possibility of philosophic rhetoric. Phaedrus doesn't appear to be very bright, and has an awful sense of rhetorical quality, since he quotes from memory a really bad speech by the orator Lysias. But one soon learns that this bad speech is simply a setup, an easy act for Socrates to follow with a much better speech of his own, and following that with a still better speech, one of the finest in all the *Dialogues* of Plato.

Beyond that, the only remarkable thing about Phaedrus is his personality. Plato often names Socrates' foils for characteristics of their personality. A young, overtalkative, innocent and good-natured foil in the *Gorgias* is named Polus, which is Greek for "colt." Phaedrus' personality is different from this. He is unallied to any particular group. He prefers the solitude of the country to the city. He is aggressive to the point of being dangerous. At one point he threatens Socrates with violence. *Phaedrus,* in Greek, means "wolf." In this dialogue he is carried away by Socrates' discourse on love and is tamed.

Our Phaedrus reads the dialogue and is tremendously impressed by the magnificent poetic imagery. But he's not tamed by it because he also smells it in a faint odor of hypocrisy. The speech is not an end in itself, but is being used to condemn that same affective domain of understanding it makes its rhetorical appeal to. The passions are characterized as the destroyer of understanding, and Phaedrus wonders if this is where the condemnation of the passions so deeply buried in Western thought got its start. Probably not. The tension between ancient Greek thought and emotion is described elsewhere as basic to Greek makeup and culture. Interesting though.

The next week the Professor of Philosophy again does not appear, and Phaedrus uses the time to catch up on his work at the University of Illinois.

The next week, in the University of Chicago bookstore across the street from where he is about to attend class, Phaedrus sees two dark eyes that stare at him steadily through a shelf of books. When the face appears he recognizes it as the face of the innocent student who had been verbally beaten up earlier in the quarter and had disappeared. The face looks as though the student knows something Phaedrus doesn't know. Phaedrus walks over to talk, but the face retreats and goes out the door, leaving Phaedrus puzzled. And on edge. Perhaps he's just fatigued and jumpy. The exhaustion of teaching at Navy Pier on top of the effort to outflank the whole body of Western academic thought at the University of Chicago is forcing him to work and study twenty hours a day with inadequate attention to food or exercise. It could be just fatigue that makes him think something is odd about that face.

But when he walks across the street to the class, the face follows about twenty paces behind. Something is up.

Phaedrus enters the classroom and waits. Soon, there comes the student again, back into the room after all these weeks. He can't expect to get credit *now.* The student looks at Phaedrus with a half-smile. He's smiling at something, all right.

At the doorway there are some footsteps, and then Phaedrus suddenly *knows*—and his legs turn rubbery and his hands start to shake. Smiling benignly in the doorway, stands none other than the Chairman for the

Committee on Analysis of Ideas and Study of Methods at the University of Chicago. He is taking over the class.

This is it. This is where they throw Phaedrus out the front door.

Courtly, grand, with imperial magnanimity the Chairman stands in the doorway for a moment, then talks to a student who seems to know him. He smiles, while looking away from the student, around the classroom, as if to find another face that is familiar to him, nods and then chuckles a little, waiting for the bell to ring.

That's why that kid is here. They've explained to him why they accidentally beat him up, and just to show what good guys they are they're going to let him have a ringside seat while they beat up Phaedrus.

How are they going to do it? Phaedrus already knows. First they are going to destroy his status dialectically in front of the class by showing how little he knows about Plato and Aristotle. That won't be any trouble. Obviously they know a hundred times more about Plato and Aristotle than he ever will. They've been at it all their lives.

Then, when they have thoroughly cut him up dialectically, they will suggest that he either shape up or get out. Then they are going to ask some more questions, and he won't know the answers to those either. Then they are going to suggest that his performance is so abominable that he not bother to attend, but leave the class right now. There are variations possible but this is the basic format. It's so easy.

Well, he has learned a lot, which is what he has come for. He can do his thesis in some other way. With that thought the rubbery feeling leaves him and he calms down.

Phaedrus has grown a beard since the Chairman last saw him, and so is still unidentified. No long advantage. The Chairman will locate him soon enough.

The Chairman lays his coat down carefully, takes a chair on the opposite side of the large round table, sits, and then brings out an old pipe and stuffs it for what must be nearly a half a minute. One can see he has done this many times before.

In a moment of attention to the class he studies face with a smiling hypnotic gaze, sensing the mood, but feeling it is not just right. He stuffs the pipe some more, but without hurry.

Soon the moment arrives, he lights the pipe, and before long there is in the classroom an odor of smoke.

At last he speaks:

"It is my understanding," he says, "that today we are to begin discussion of the immortal *Phaedrus.*" He looks at each student separately. "Is that correct?"

Members of the class assure him timidly that it is. His persona is overwhelming.

The Chairman then apologizes for the absence of the previous Professor, and describes the format of what will follow. Since he already knows the dialogue himself he will elicit from the class answers that will show how well they have studied it.

That's the best way to do it, Phaedrus thinks. That way one can learn to know the individual students. Fortunately Phaedrus has studied the dialogue so carefully it is almost memorized.

The Chairman is right. It is an immortal dialogue, strange and puzzling at first, but then hitting you harder and harder, like truth itself. What Phaedrus has been talking about as Quality, Socrates appears to have described as the soul, self-moving, the source of all things. There is no contradiction. There never really can be between the core terms of monistic philosophies. The One in India has got to be the same as the One in Greece. If it's not, you've got two. The only disagreements among the monists concern the attributes of the One, not the One itself. Since the One is the source of all things and includes all things in it, it cannot be defined in terms of those things, since no matter what thing you use to define it, the thing will always describe something less than the One itself. The One can only be described allegorically, through the use of analogy, of figures of imagination and speech. Socrates chooses a heaven-and-earth analogy, showing how individuals are drawn toward the One by a chariot drawn by two horses. . . .

But the Chairman now directs a question to the student next to Phaedrus. He is baiting him a little, provoking him to attack.

The student, whose identity is mistaken, doesn't attack, and the Chairman with great disgust and frustration finally dismisses him with a rebuke that he should have read the material better.

Phaedrus' turn. He has calmed down tremendously. He must now explain the dialogue.

"If I may be permitted to begin again in my own way," he says, partly to conceal the fact that he didn't hear what the previous student said.

The Chairman, seeing this as a further rebuke to the student next to him, smiles and says contemptuously it is certainly a good idea.

Phaedrus proceeds. "I believe that in this dialogue the person of Phaedrus is characterized as a *wolf.*"

He has delivered this quite loudly, with a flash of anger, and the Chairman almost jumps. Score!

"Yes," the Chairman says, and a gleam in his eye shows he now recognizes who his bearded assailant is. "*Phaedrus* in Greek does mean 'wolf.' That's a very acute observation." He begins to recover his composure. "Proceed."

"Phaedrus meets Socrates, *who knows only the ways of the city,* and leads him into the country, whereupon he begins to recite a speech of the orator, Lysias, whom he admires. Socrates asks him to read it and Phaedrus does."

"Stop!" says the Chairman, who has now completely recovered his composure. "You are giving us the plot, not the dialogue." He calls on the next student.

None of the students seems to know to the Chairman's satisfaction what the dialogue is about. And so with mock sadness he says they must all read more thoroughly but this time he will help them by taking on the burden of explaining the dialogue himself. This provides an overwhelming relief to the tension he has so carefully built up and the entire class is in the palm of his hand.

The Chairman proceeds to reveal the meaning of the dialogue with complete attention. Phaedrus listens with deep engagement.

After a time something begins to disengage him a little. A false note of some kind has crept in. At first he doesn't see what it is, but then he becomes aware that the Chairman has completely bypassed Socrates' description of the One and has jumped ahead to the allegory of the chariot and the horses.

In this allegory the seeker, trying to reach the One, is drawn by two horses, one white and noble and temperate, and the other surly, stubborn, passionate and black. The one is forever aiding him in his upward journey to the portals of heaven, the other is forever confounding him. The Chairman has not stated it yet, but he is at the point at which he must now announce that the white horse is temperate reason, the black horse is dark passion, emotion. He is at the point at which these must be described, but the false note suddenly becomes a chorus.

He backs up and restates that "Now Socrates has sworn to the Gods that he is telling the Truth. He has taken an oath to speak the Truth, and if what follows is not the Truth he has forfeited his own soul."

TRAP! He's *using* the dialogue to prove the holiness of reason! Once that's established he can move down into inquiries of what reason is, and then, lo and behold, there we are in Aristotle's domain again!

Phaedrus raises his hand, palm flat out, elbow on the table. Where before this hand was shaking, it is now deadly calm. Phaedrus senses that he now is formally signing his own death warrant here, but knows he will sign another kind of death warrant if he takes his hand down.

The Chairman sees the hand, is surprised and disturbed by it, but acknowledges it. Then the message is delivered.

Phaedrus says, "All this is just an analogy."

Silence. And then confusion appears on the Chairman's face. "What?" he says. The spell of his performance is broken.

"This entire description of the chariot and horses is just an analogy."

"What?" he says again, then loudly, "It is the *truth*! Socrates has sworn to the Gods that it is the truth!"

Phaedrus replies, "Socrates himself says it is an analogy."

"If you will read the dialogue you will find that Socrates specifically states it is the Truth!"

"Yes, but *prior* to that . . . in, I believe, two paragraphs . . . he has stated that it is an *analogy*."

The text is on the table to consult but the Chairman has enough sense not to consult it. If he does and Phaedrus is right, his classroom face is completely demolished. He has told the class no one has read the book thoroughly.

Rhetoric, 1; Dialectic, 0.

Fantastic, Phaedrus thinks, that he should have remembered that. It just demolishes the whole dialectical position. That may just be the whole show right there. Of course it's an analogy. Everything is an analogy. But the dialecticians don't know that. That's why the Chairman missed that statement of Socrates. Phaedrus has caught it and remembered it, because if Socrates hadn't stated it he wouldn't have been telling the "Truth."

No one sees it yet, but they will soon enough. The Chairman of the Committee on Analysis of Ideas and Study of Methods has just been shot down in his own classroom.

Now he is speechless. He can't think of a word to say. The silence which so built his image at the beginning of the class is now destroying it. He doesn't understand from where the shot has come. He has never confronted a living Sophist. Only dead ones.

Now he tries to grasp on to something, but there is nothing to grasp on to. His own momentum carries him forward into the abyss, and when he finally finds words they are the words of another kind of person; a schoolboy who has forgotten his lesson, has gotten it wrong, but would like our indulgence anyway.

He tries to bluff the class with the statement he made before that no one has studied very well, but the student to Phaedrus' right shakes his head at him. Obviously someone has.

The Chairman falters and hesitates, acts afraid of his class and does not really engage them. Phaedrus wonders what the consequences of this will be.

Then he sees a bad thing happen. The beat-up innocent student who has watched him earlier now is no longer so innocent. He is sneering at the Chairman and asking him sarcastic and insinuating questions. The Chairman, already crippled, is now being killed . . . but then Phaedrus realizes this was what was intended for himself.

He can't feel sorry, just disgusted. When a shepherd goes to kill a wolf, and takes his dog to see the sport, he should take care to avoid mistakes. The dog has certain relationships to the wolf the shepherd may have forgotten.

A girl rescues the Chairman by asking easy questions. He receives the questions with gratitude, answers each at great length and slowly recovers himself.

Then the question is asked him, "What is dialectic?"

He thinks about it, and then, by God, turns to Phaedrus and asks if he would care to answer.

"You mean my *personal* opinion?" Phaedrus asks.

"No . . . let us say, Aristotle's opinion."

No subtleties now. He is just going to get Phaedrus on his own territory and let him have it.

"As best I know . . ." Phaedrus says, and pauses.

"Yes?" The Chairman is all smiles. Everything is all set.

"As best I know, Aristotle's opinion is that dialectic comes *before* everything else."

The Chairman's expression goes from unction to shock to rage in one-half second flat. It *does*! his face shouts, but he never says it. The trapper trapped again. He can't kill Phaedrus on a statement taken from his own article in the *Encyclopaedia Britannica*.

Rhetoric, 2; Dialectic, 0.

"And from the dialectic come the forms," Phaedrus continues, "and from . . ." But the Chairman cuts it off. He sees it cannot go his way and dismisses it.

He shouldn't have cut it off, Phaedrus thinks to himself. Were he a real Truth-seeker and not a propagandist for a particular point of view he would not. He might learn something. Once it's stated that "the dialectic comes before anything else," this statement itself becomes a dialectical entity, subject to dialectical question.

Phaedrus would have asked, What evidence do we have that the dialectical question-and-answer method of arriving at truth comes before anything else? We have none whatsoever. And when the statement is isolated and itself subject to scrutiny it becomes patently ridiculous. Here is this dialectic, like Newton's law of gravity, just sitting by itself in the middle of nowhere, giving birth to the universe, hey? It's asinine.

Dialectic, which is the parent of logic, came itself

from rhetoric. Rhetoric is in turn the child of the myths and poetry of ancient Greece. That is so historically, and that is so by any application of common sense. The poetry and the myths are the response of a prehistoric people to the universe around them made on the basis of Quality. It is Quality, not dialectic, which is the generator of everything we know.

The class ends, the Chairman stands by the door answering questions, and Phaedrus almost goes up to say something but does not. A lifetime of blows tends to make a person unenthusiastic about any unnecessary interchange that might lead to more. Nothing friendly has been said or even hinted at and much hostility has been shown.

Phaedrus the wolf. It fits. Walking back to his apartment with light steps he sees it fits more and more. He wouldn't be happy if they were overjoyed with the thesis. Hostility is really his element. It really is. Phaedrus the wolf, yes, down from the mountains to prey upon the poor innocent citizens of this intellectual community. It fits all right.

The Church of Reason, like all institutions of the System, is based not on individual strength but upon individual weakness. What's really demanded in the Church of Reason is not ability, but *in*ability. Then you are considered teachable. A truly able person is always a threat. Phaedrus sees that he has thrown away a chance to integrate himself into the organization by submitting to whatever Aristotelian thing he is supposed to submit to. But that kind of opportunity seems hardly worth the bowing and scraping and intellectual prostration necessary to maintain it. It is a low-quality form of life.

For him Quality is better seen up at the timberline than here obscured by smoky windows and oceans of words, and he sees that what he is talking about can never really be accepted here because to see it one has to be free from social authority and this is an institution of social authority. Quality for sheep is what the shepherd says. And if you take a sheep and put it up at the timberline at night when the wind is roaring, that sheep will be panicked half to death and will call and call until the shepherd comes, or comes the wolf.

He makes one last attempt somehow to be nice at the next session of the class but the Chairman isn't having any. Phaedrus asks him to explain a point, saying he hasn't been able to understand it. He has, but thinks it would be nice to defer a little.

The answer is "Maybe you got tired!" delivered as scathingly as possible; but it doesn't scathe. The Chairman is simply condemning in Phaedrus that which he most fears in himself. As the class goes on Phaedrus sits staring out the window feeling sorry for this old shepherd and his classroom sheep and dogs and sorry for himself that he will never be one of them. Then, when the bell rings, he leaves forever.

The classes at Navy Pier by contrast are going like wildfire, the students now listening intently to this strange, bearded figure from the mountains who is telling them there was such a thing as Quality in this universe and they know what it is. They don't know what to make of it, are unsure, some of them afraid of him. They can see he is somehow dangerous, but all are fascinated and want to hear more.

But Phaedrus is no shepherd either and the strain of behaving like one is killing him. A strange thing that has always occurred in classes occurs again, when the unruly and wild students in the back rows have always empathized with him and been his favorites, while the more sheepish and obedient students in the front rows have always been terrorized by him and are because of this objects of his contempt, even though in the end the sheep have passed and his unruly friends in the back rows have not. And Phaedrus sees, though he does not want to admit it to himself even now, he sees intuitively nevertheless that his days as a shepherd are coming to an end too. And he wonders more and more what is going to happen next.

He has always feared the silence in the classroom, the sort that has destroyed the Chairman. It is not his nature to talk and talk and talk for hours on end and it exhausts him to do this, and now, having nothing left to turn upon, he turns upon this fear.

He comes to the classroom, the bell rings, and Phaedrus sits there and does not talk. For the entire hour he is silent. Some of the students challenge him a little to wake him up, but then are silent. Others are going straight out of their minds with internal panic. At the end of the hour the whole class literally breaks and runs for the door. Then he goes to his next class and the same thing happens. And the next class, and the next. Then Phaedrus goes home. And he wonders more and more what is going to happen next.

Thanksgiving comes.

His four hours of sleep have dwindled down to two and then to nothing. It is all over. He will not be going back to the study of Aristotelian rhetoric. Neither will he return to the teaching of that subject. It is over. He begins to walk the streets, his mind spinning.

The city closes in on him now, and in his strange perspective it becomes the antithesis of what he believes. The citadel not of Quality, the citadel of form and substance. Substance in the form of steel sheets and girders, substance in the form of concrete piers and roads, in the form of brick, of asphalt, of auto parts, old radios, and rails, dead carcasses of animals that once grazed the prairies. Form and substance without Quality. That is the soul of this place. Blind, huge, sinister and inhuman: seen by the light of fire flaring upward in the night from the blast furnaces in the south, through heavy coal smoke deeper and denser into the neon of BEER and PIZZA and LAUNDROMAT signs and unknown and meaningless signs along meaningless straight streets going off into other straight streets forever.

If it was all bricks and concrete, pure forms of substance, clearly and openly, he might survive. It is the little, pathetic attempts at Quality that kill. The plaster false fireplace in the apartment, shaped and waiting to contain a flame that can never exist. Or the hedge in front of the apartment building with a few square feet of grass behind it. A few square feet of grass, after Montana. If they just left out the hedge and grass it would be all right. Now it serves only to draw attention to what has been lost.

Along the streets that lead away from the apartment he can never see anything through the concrete and brick and neon but he knows that buried within it are grotesque, twisted souls forever trying the manners that will convince themselves they possess Quality, learning strange poses of style and glamour vended by dream magazines and other mass media, and paid for by the vendors of substance. He thinks of them at night alone with their advertised glamorous shoes and stockings and underclothes off, staring through the sooty windows at the grotesque shells revealed beyond them, when the poses weaken and the truth creeps in, the only truth that exists here, crying to heaven, God, there is nothing here but dead neon and cement and brick.

His time consciousness begins to go. Sometimes his thoughts race on and on at a speed seeming to approach that of light. But when he tries to make decisions relating to his surroundings, it seems to take whole minutes for a single thought to emerge. A single thought begins to grow in his mind, extracted from something he read in the dialogue *Phaedrus*.

"And what is written well and what is written badly—need we ask Lysias or any other poet or orator who ever wrote or will write either a political or other work, in meter or out of meter, poet or prose writer, to teach us this?"

What is good, Phaedrus, and what is not good—need we ask anyone to tell us these things?

It is what he was saying months before in the classroom in Montana, a message Plato and every dialectician since him had missed, since they all sought to define the Good in its intellectual relation to things. But what he sees now is how far he has come from that. He is doing the same bad things himself. His original goal was to keep Quality undefined, but in the process of battling against the dialecticians he has made statements, and each statement has been a brick in a wall of definition he himself has been building around Quality. Any attempt to develop an organized reason around an undefined quality defeats its own purpose. The organization of the reason itself defeats the quality. Everything he has been doing has been a fool's mission to begin with.

On the third day he turns a corner at an intersection of unknown streets and his vision blanks out. When it returns he is lying on the sidewalk, people moving around him as if he were not there. He gets up wearily and mercilessly drives his thoughts to remember the way back to the apartment. They are slowing down. Slowing down. This is about the time he and Chris try to find the sellers of bunk beds for the children to sleep in. After that he does not leave the apartment.

He stares at the wall in a cross-legged position upon a quilted blanket on the floor of a bedless bedroom. All bridges have been burned. There is no way back. And now there is no way forward either.

For three days and three nights, Phaedrus stares at the wall of the bedroom, his thoughts moving neither forward nor backward, staying only at the instant. His wife asks if he is sick, and he does not answer. His wife becomes angry, but Phaedrus listens without responding. He is aware of what she says but is no longer able to feel any urgency about it. Not only are his thoughts slowing down, but his desires too. And they slow and slow, as if gaining an imponderable mass. So heavy, so tired, but no sleep comes. He feels like a giant, a million miles tall. He feels himself extending into the universe with no limit.

He begins to discard things, encumbrances that he has carried with him all his life. He tells his wife to leave with the children, to consider themselves separated. Fear of loathsomeness and shame disappear when his urine flows not deliberately but naturally on the floor of the room. Fear of pain, the pain of the martyrs is overcome when cigarettes burn not deliberately but naturally down into his fingers until they are extinguished by blisters formed by their own heat. His wife sees his injured hands and the urine on the floor and calls for help.

But before help comes, slowly, imperceptibly at first, the entire consciousness of Phaedrus begins to come apart . . . to dissolve and fade away. Then gradually he no longer wonders what will happen next. He knows what will happen next, and tears flow for his family and for himself and for this world. A fragment comes and lingers from an old Christian hymn, "You've got to cross that lonesome valley." It carries him forward. "You've got to cross it by yourself." It seems a Western hymn that belongs out in Montana.

"No one else can cross it for you," it says. It seems to suggest something beyond. "You've got to cross it by yourself."

He crosses a lonesome valley, out of the mythos, and emerges as if from a dream, seeing that his whole consciousness, the mythos, has been a dream and no one's dream but his own, a dream he must now sustain of his own efforts. Then even "he" disappears and only the dream of himself remains with himself in it.

And the Quality, the *aretê* he has fought so hard for, has sacrificed for, has *never* betrayed, but in all that time has never once understood, now makes itself clear to him and his soul is at rest.

*   *   *

The cars are thinned out to almost none, and the road is so black it seems as though the headlight can barely fight its way through the rain to reach it. Murderous. Anything can happen—a sudden rut, an oil slick, a dead animal. . . . But if you go too slow they'll kill you from behind. I don't know why we still go on in this. We should have stopped long ago. I don't know what I'm doing anymore. I was looking for some sign of a motel, I guess, but not thinking about it and missing them. If we keep on like this they'll all close.

We take the next exit from the freeway, hoping it will lead somewhere, and soon are on bumpy blacktop with ruts and loose gravel. I go slowly. Streetlamps overhead throw swinging arcs of sodium light through the sheets of rain. We pass from light into shadow into light into shadow again without a single sign of welcome anywhere. A sign announces "STOP" to our left, but does not tell which way to turn. One way looks as dark as the other. We could go endlessly through these streets and not find anything, and now not even find the freeway again.

"Where are we?" Chris shouts.

"I don't know." My mind has become tired and slow. I can't seem to think of the right answer . . . or what to do next.

Now I see ahead a white glow and bright sign of a filling station far down the street.

It's open. We pull up and go inside. The attendant, who looks Chris's age, watches us strangely. He doesn't know of any motel. I go to the telephone directory, find some and tell him the street addresses, and he tries to give directions but they're poor. I call the motel he says is closest, make a reservation and confirm the directions.

In the rain and the dark streets, even with directions, we almost miss it. They have turned the light out, and when I register nothing is said.

The room is a remnant of the bleakness of the thirties, sordid, homemade by a person who didn't know carpentry, but it's dry and has a heater and beds and that's all we want. I turn on the heater and we sit before it and soon the chills and shivers and damp start to leave our bones.

Chris doesn't look up, just stares into the grille of the wall heater. Then, after a while, he says, "When are we going back home?"

Failure.

"When we get to San Francisco," I say. "Why?"

"I'm so tired of just sitting and . . ." His voice has trailed off.

"And what?"

"And . . . I don't know. Just sitting . . . like we're not really going anyplace."

"Where should we go?"

"I don't know. How should I know?"

"I don't know either," I say.

"Well, why *don't* you!" he says. He begins to cry.

"What's the matter, Chris?" I ask.

He doesn't answer. Then he puts his head in his hands and rocks back and forth. The way he does it gives me an eerie feeling. After a while he stops and says, "When I was little it was different."

"How?"

"I don't know. We always *did* things. That I wanted to do. Now I don't want to do *any*thing."

He continues to rock back and forth in that eerie way, with his face in his hands, and I don't know what to do. It's a strange, unworldly rocking motion, a fetal self-enclosure that seems to shut me out, to shut everything out. A return to somewhere that I don't know about . . . the bottom of the ocean.

Now I know where I have seen it before, on the floor of the hospital.

I don't know of anything to do.

After a while we get in our beds and I try to sleep.

Then I ask Chris, "Was it better before we left Chicago?"

"Yes."

"How? What do you remember?"

"That was fun."

"Fun?"

"Yes," he says, and is quiet. Then he says, "Remember the time we went to look for beds?"

"That was *fun*?"

"Sure," he says, and is quiet for a long time. Then he says, "Don't you remember? You made me find all the directions home. . . . You used to play games with us. You used to tell us all kinds of stories and we'd go on rides to do things and now you don't do anything."

"Yes, I do."

"No, you *don't*! You just sit and stare and you don't *do anything*!" I hear him crying again.

Outside the rain comes in gusts against the window, and I feel a kind of heavy pressure bear down on me. He's crying for *him*. It's *him* he misses. That's what the dream is about. In the dream . . .

*   *   *

For what seems like a long time I continue to listen to the cricking sound of the wall heater and the wind and the rain against the roof and window. Then the rain dies away and there is nothing left but a few drops of water from the trees moving in an occasional gust of wind.

31

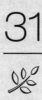

In the morning I'm stopped by the appearance of a green slug on the ground. It's about six inches long, three-quarters inch wide and soft and almost rubbery and covered with slime like some internal organ of an animal.

All around me it's damp and wet and foggy and cold, but clear enough to see that the motel we have stopped in is on a slope with apple trees down below and grass and small weeds under them covered with dew or just rain that hasn't run off. I see another slug and then another—my God, the place is crawling with them.

When Chris comes out I show one to him. It moves slowly like a snail across a leaf. He has no comment.

We leave and breakfast in a town off the road called Weott, where I see he's still in a distant mood. It's a kind of looking-away mood and a not-talking mood, and I leave him alone.

Farther on at Leggett we see a tourist duck pond and

we buy Cracker Jacks and throw them to the ducks and he does this in the most unhappy way I have ever seen. Then we pass into some of the twisting coastal range road and suddenly enter heavy fog. Then the temperature drops and I know we're back at the ocean again.

When the fog lifts we can see the ocean from a high cliff, far out and so blue and so distant. As we ride I become colder, deep cold.

We stop and I get out the jacket and put it on. I see Chris go very close to the edge of the cliff. It's at least one hundred feet to the rocks below. Way too close!

"CHRIS!" I holler. He doesn't answer.

I go up, swiftly grab his shirt and pull him back. "Don't do that," I say.

He looks at me with a strange squint.

I get out extra clothes for him and hand them to him. He takes them but he dawdles and doesn't put them on.

There's no sense hurrying him. In this mood if he wants to wait, he can.

He waits and waits. Ten minutes, then fifteen minutes pass.

We're going to have a waiting contest.

After thirty minutes of cold winds off the ocean he asks, "Which way are we going?"

"South, now, along the coast."

"Let's go back."

"Where?"

"To where it's warmer."

That would add another hundred miles. "We have to go south now," I say.

"Why?"

"Because it would add too many miles going back."

"Let's go back."

"No. Get your warm clothes on."

He doesn't and just sits there on the ground.

After another fifteen minutes he says, "Let's go back."

"Chris, you're not running the cycle. I'm running it. We're going south."

"Why?"

"Because it's too far and because I've said so."

"Well, why don't we just go back?"

Anger reaches me. "You don't really want to know, do you?"

"I want to go back. Just tell me why we can't go back."

I'm hanging on to my temper now. "What you really want isn't to go back. What you really want is just to get me angry, Chris. If you keep it up you'll succeed!"

Flash of fear. That's what he wanted. He wants to hate me. Because I'm not *him*.

He looks down at the ground bitterly, and puts his warm clothes on. Then we're back on the machine and moving down the coast again.

I can imitate the father he's supposed to have, but subconsciously, at the Quality level, he sees through it and knows his real father isn't here. In all this Chautauqua talk there's been more than a touch of hypocrisy. Advice is given again and again to eliminate subject-object duality, when the biggest duality of all, the duality between me and him, remains unfaced. A mind divided against itself.

But who did it? *I* didn't do it. And there's no way now of undoing it. . . . I keep wondering how far it is to the bottom of that ocean out there. . . .

*   *   *

What I am is a heretic who's recanted, and thereby in everyone's eyes saved his soul. Everyone's eyes but one, who knows deep down inside that all he has saved is his skin.

I survive mainly by pleasing others. You do that to get out. To get out you figure out what they want you to say and then you say it with as much skill and originality as possible and then, if they're convinced, you get out. If I hadn't turned on him I'd still be there, but he was true to what he believed right to the end. That's the difference between us, and Chris knows it. And that's the reason why sometimes I feel he's the reality and I'm the ghost.

We're on the Mendocino County coast now, and it's all wild and beautiful and open here. The hills are mostly grass but in the lee of rocks and folds in the hills are strange flowing shrubs sculptured by the upsweep of winds from the ocean. We pass some old wooden fences, weathered grey. In the distance is an old weathered and grey farmhouse. How could anyone farm here? The fence is broken in many places. Poor.

Where the road drops down from the high cliffs to the beach we stop to rest. When I turn the engine off Chris says, "What are we stopping here for?"

"I'm tired."

"Well, I'm not. Let's keep going." He's angry still. I'm angry too.

"Just go over on the beach there and run around in circles until I'm done resting," I say.

"Let's keep going," he says, but I walk away and ignore it. He sits on the curb by the motorcycle.

The ocean smell of rotting organic matter is heavy here and the cold wind doesn't allow much rest. But I find a large cluster of grey rocks where the wind is still and the heat of the sun can still be felt and enjoyed. I concentrate on the warmth of the sunlight and am grateful for what little there is.

We ride again and what comes to me now is the realization that he's another Phaedrus, thinking the way he used to and acting the same way he used to, looking for trouble, being driven by forces he's only dimly aware of and doesn't understand. The questions . . . the same questions . . . he's got to know everything.

And if he doesn't get the answer he just drives and drives until he gets one and that leads to another question and he drives and drives for the answer to that . . . endlessly pursuing questions, never seeing, never understanding that the questions will never end. Something is missing and he knows it and will kill himself trying to find it.

We round a sharp turn up an overhanging cliff. The ocean stretches forever, cold and blue out there, and produces a strange sense of despair. Coastal people never really know what the ocean symbolizes to landlocked inland people—what a great distant dream it is, present but unseen in the deepest levels of subconsciousness, and when they arrive at the ocean and the conscious images are compared with the subconscious dream there is a sense of defeat at having come so far to be so stopped by a mystery that can never be fathomed. The source of it all.

A long time later we come to a town where a luminous haze which has seemed so natural over the ocean is now seen in the streets of the town, giving them a certain aura, a hazy sunny radiance that makes everything look nostalgic, as if remembered from years before.

We stop in a crowded restaurant and find the last remaining empty table by a window overlooking the radiant street. Chris looks down and doesn't talk. Maybe, in some way, he senses that we haven't much farther to go.

"I'm not hungry," he says.

"You don't mind waiting while I eat?"

"Let's keep going. I'm not hungry."

"Well, I am."

"Well, I'm not. My stomach hurts." The old symptom.

I eat my lunch amid the conversation and clink of plates and spoons from the other tables and out the window watch a bicycle and rider go by. I feel like somehow we have arrived at the end of the world.

I look up and see Chris is crying.

"Now what?" I say.

"My stomach. It's hurting."

"Is that all?"

"No. I just *hate* everything . . . I'm sorry I came . . . I hate this trip . . . I thought this was going to be fun, and it isn't any fun . . . I'm sorry I came." He is a truth-teller, like Phaedrus. And like Phaedrus he looks at me now with more and more hatred. The time has come.

"I've been thinking, Chris, of putting you on the bus here with a ticket for home."

His face has no expression on it, then surprise mixed with dismay.

I add, "I'll go on myself with the motorcycle and see you in a week or two. There's no sense forcing you to continue on a vacation you hate."

Now it's my turn to be surprised. His expression isn't relieved at all. The dismay gets worse and he looks down and says nothing.

He seems caught off balance now, and frightened.

He looks up. "Where would I stay?"

"Well, you can't stay at our house now, because other people are there. You can stay with Grandma and Grandpa."

"I don't want to stay with them."

"You can stay with your aunt."

"She doesn't like me. I don't like her."

"You can stay with your other grandma and grandpa."

"I don't want to stay there either."

I name some others but he shakes his head.

"Well, *who* then?"

"I don't know."

"Chris, I think you can see for yourself what the problem is. You don't want to be on this trip. You hate it. Yet you don't want to stay with anyone or go anywhere else. All these people I've mentioned you either don't like or they don't like you."

He's silent but tears now form.

A woman at another table is looking at me angrily. She opens her mouth as if about to say something. I turn a heavy gaze on her for a long time until she closes her mouth and goes back to eating.

Now Chris is crying hard and others look over from the other tables.

"Let's go for a walk," I say, and get up without waiting for the check.

At the cash register the waitress says, "I'm sorry the boy isn't feeling good." I nod, pay, and we're outside.

I look for a bench somewhere in the luminous haze but there is none. Instead we climb on the cycle and go slowly south looking for a restful place to pull off.

The road leads out to the ocean again where it climbs to a high point that apparently juts out into the ocean but now is surrounded by banks of fog. For a moment I see a distant break in the fog where some people rest in the sand, but soon the fog rolls in and the people are obscured.

I look at Chris and see a puzzled, empty look in his eyes, but as soon as I ask him to sit down some of the anger and hatred of this morning reappear.

"Why?" he asks.

"I think it's time we should talk."

"Well, talk," he says. All the old belligerence is back. It's the "kind father" image he can't stand. He knows the "niceness" is false.

"What about the future?" I say. Stupid thing to ask.

"What about it?" he says.

"I was going to ask what you planned to do about the future."

"I'm going to let it be." Contempt shows now.

The fog opens for a moment, revealing the cliff we are on, then closes again, and a sense of inevitability about what is happening comes over me. I'm being pushed toward something and the objects in the corner of the eye and the objects in the center of the vision are all of equal intensity now, all together in one, and I say, "Chris, I think it's time to talk about some things you don't know about."

He listens a little. He senses something is coming.

"Chris, you're looking at a father who was insane for a long time, and is close to it again."

And not just close anymore. It's here. The bottom of the ocean.

"I'm sending you home not because I'm angry with you but because I'm afraid of what can happen if I continue to take responsibility for you."

His face doesn't show any change of expression. He doesn't understand yet what I'm saying.

"So this is going to be good-bye, Chris, and I'm not sure we'll see each other anymore."

That's it. It's done. And now the rest will follow naturally.

He looks at me so strangely. I think he still doesn't understand. That gaze . . . I've seen it somewhere . . . somewhere . . . somewhere. . . .

In the fog of an early morning in the marshes there was a small duck, a teal that gazed like this . . . I'd winged it and now it couldn't fly and I'd run up on it and seized it by the neck and before killing it had stopped and from some sense of the mystery of the universe had stared into its eyes, and they gazed like this . . . so calm and uncomprehending . . . and yet so aware. Then I closed my hands around its eyes and twisted the neck until it broke and I felt the snap between my fingers.

Then I opened my hand. The eyes still gazed at me but they stared into nothing and no longer followed my movements.

"Chris, they're saying it about you."

He gazes at me.

"That all these troubles are in your mind."

He shakes his head no.

"They seem real and feel real but they aren't."

His eyes become wide. He continues to shake his head no, but comprehension overtakes him.

"Things have gone from bad to worse. Trouble in school, trouble with the neighbors, trouble with your family, trouble with your friends . . . trouble everywhere you turn. Chris, I was the only one holding them all back, saying, 'He's all right,' and now there won't be anyone. Do you understand?"

He stares stunned. His eyes still track but they begin to falter. I'm not giving him strength. I never have been. I'm killing him.

"It's not your fault, Chris. It never has been. Please understand that."

His gaze fails in a sudden inward flash. Then his eyes close and a strange cry comes from his mouth, a wail like the sound of something far away. He turns and stumbles on the ground then falls, doubles up and kneels and rocks back and forth, head on the ground. A faint misty wind blows in the grass around him. A seagull alights nearby.

Through the fog I hear the whine of gears of a truck and am terrified by it.

"You have to get up, Chris."

The wail is high-pitched and inhuman, like a siren in the distance.

"You must get up!"

He continues to rock and wail on the ground.

I don't know what to do now. I have no idea what to do. It's all over. I want to run for the cliff, but fight that. I have to get him on the bus, and then the cliff will be all right.

"Everything is all right now, Chris."

That's not my voice.

"I haven't forgotten you."

Chris's rocking stops.

"How could I forget you?"

Chris raises his head and looks at me. A film he has always looked through at me disappears for a moment and then returns.

"We'll be together now."

The whine of the truck is upon us.

"Now get up!"

Chris slowly sits up and stares at me. The truck arrives, stops, and the driver looks out to see if we need a ride. I shake my head no and wave him on. He nods, puts the truck in gear, and it whines off through the mist again and there is only Chris and me.

I put my jacket around him. His head is buried again between his knees and he cries now, but it is a low-pitched human wail, not the strange cry of before. My hands are wet and I feel that my forehead is wet too.

After a while he wails, "Why did you leave us?"

"When?"

"At the hospital!"

"There was no choice. The police prevented it."

"Wouldn't they let you out?"

"No."

"Well then, why wouldn't you open the door?"

"What door?"

"The glass door!"

A kind of slow electric shock passes through me. What glass door is he talking about?

"Don't you remember?" he says. "We were standing on one side and you were on the other side and Mom was crying."

I've never told him about that dream. How could he know about that? Oh, no.

We're in another dream. That's why my voice sounds so strange.

"I couldn't open that door. They told me not to open it. I had to do everything they said."

"I thought you didn't want to see us," Chris says. He looks down.

The looks of terror in his eyes all these years.

Now I see the door. It is in a hospital.

This is the last time I will see them. I am Phaedrus, that is who I am, and they are going to destroy me for speaking the Truth.

It has all come together.

Chris cries softly now. Cries and cries and cries. The wind from the ocean blows through the tall stems of grass all around us and the fog begins to lift.

"Don't cry, Chris. Crying is just for children."

After a long time I give him a rag to wipe his face with. We gather up our stuff and pack it on the motorcycle. Now the fog suddenly lifts and I see the sun on his face makes his expression open in a way I've never seen it before. He puts on his helmet, tightens the strap, then looks up.

"Were you really insane?"

Why should he ask that?

"No!"

Astonishment hits. But Chris's eyes sparkle.

"I knew it," he says.

Then he climbs on the cycle and we are off.

32

As we ride now through coastal manzanita and waxen-leafed shrubs, Chris's expression comes to mind. "I knew it," he said.

The cycle swings into each curve effortlessly, banking so that our weight is always down through the machine no matter what its angle is with the ground. The way is full of flowers and surprise views, tight turns one after another so that the whole world rolls and pirouettes and rises and falls away.

"I knew it," he said. It comes back now as one of those little facts tugging at the end of a line, saying it's not as small as I think it is. It's been in his mind for a long time. Years. All the problems he's given become more understandable. "I *knew* it," he said.

He must have heard something long ago, and in his childish misunderstanding gotten it all mixed up. That's what Phaedrus always said—*I* always said—years ago, and Chris must have believed it, and kept it hidden inside ever since.

We're related to each other in ways we never fully understand, maybe hardly understand at all. He was always the *real* reason for coming out of the hospital. To have let him grow up alone would have been really wrong. In the dream too he was the one who was always trying to open the door.

I haven't been carrying him at all. He's been carrying *me!*

"I *knew* it," he said. It keeps tugging on the line, saying my big problem may not be as big as I think it is, because the answer is right in front of me. For God's sake relieve him of his burden! Be one person again!

Rich air and strange perfumes from the flowers of the trees and shrubs enshroud us. Inland now the chill is gone and the heat is upon us again. It soaks through my jacket and clothes and dries out the dampness inside. The gloves which have been dark-wet have started to turn light again. It seems like I've been bone-chilled by that ocean damp for so long I've forgotten what heat is like. I begin to feel drowsy and in a small ravine ahead I see a turnoff and a picnic table. When we get to it I cut the engine and stop.

"I'm sleepy," I tell Chris. "I'm going to take a nap."

"Me too," he says.

We sleep and when we wake up I feel very rested, more rested than for a long time. I take Chris's jacket and mine and tuck them under the elastic cables holding down the pack on the cycle.

It's so hot I feel like leaving this helmet off. I remember that in this state they're not required. I fasten it around one of the cables.

"Put mine there too," Chris says.

"You need it for safety."

"You're not wearing yours."

"All right," I agree, and stow his too.

The road continues to twist and wind through the trees. It upswings around hairpins and glides into new scenes one after another around and through brush and then out into open spaces where we can see canyons stretch away below.

"Beautiful!" I holler to Chris.

"You don't need to shout," he says.

"Oh," I say, and laugh. When the helmets are off you can talk in a conversational voice. After all these days!

"Well, it's beautiful, anyway," I say.

More trees and shrubs and groves. It's getting warmer. Chris hangs onto my shoulders now and I turn a little and see that he stands up on the foot pegs.

"That's a little dangerous," I say.

"No, it isn't. I can tell."

He probably can. "Be careful anyway," I say.

After a while when we cut sharp into a hairpin under some overhanging trees he says, "Oh," and then later on, "Ah," and then, "Wow." Some of these branches over the road are hanging so low they're going to conk him on the head if he isn't careful.

"What's the matter?" I ask.

"It's so different."

"What?"

"Everything. I never could see over your shoulders before."

The sunlight makes strange and beautiful designs through the tree branches on the road. It flits light and dark into my eyes. We swing into a curve and then up into the open sunlight.

That's true. I never realized it. All this time he's been staring into my back. "What do you see?" I ask.

"It's all different."

We head into a grove again, and he says, "Don't you get scared?"

"No, you get used to it."

After a while he says, "Can I have a motorcycle when I get old enough?"

"If you take care of it."

"What do you have to do?"

"Lot's of things. You've been watching me."

"Will you show me all of them?"

"Sure."

"Is it hard?"

"Not if you have the right attitudes. It's having the right attitudes that's hard."

"Oh."

After a while I see he is sitting down again. Then he says, "Dad?"

"What?"

"Will I have the right attitudes?"

"I think so," I say. "I don't think that will be any problem at all."

And so we ride on and on, down through Ukiah, and Hopland, and Cloverdale, down into the wine country. The freeway miles seem so easy now. The engine which has carried us halfway across a continent drones on and on in its continuing oblivion to everything but its own internal forces. We pass through Asti and Santa Rosa, and Petaluma and Novato, on the freeway that grows wider and fuller now, swelling with cars and trucks and buses full of people, and soon by the road are houses and boats and the water of the Bay.

Trials never end, of course. Unhappiness and misfortune are bound to occur as long as people live, but there is a feeling now, that was not here before, and is not just on the surface of things, but penetrates all the way through: We've won it. It's going to get better now. You can sort of tell these things.

Afterword

*This book has a lot to say about Ancient Greek per-*spectives and their meaning but there is one perspective it misses. That is their view of time. They saw the future as something that came upon them from behind their backs with the past receding away before their eyes.

When you think about it, that's a more accurate metaphor than our present one. Who really *can* face the future? All you can do is project from the past, even when the past shows that such projections are often wrong. And who really can forget the past? What else is there to know?

Ten years after the publication of *Zen and the Art of Motorcycle Maintenance* the Ancient Greek perspective is certainly appropriate. What sort of future is coming up from behind I don't really know. But the past, spread out ahead, dominates everything in sight.

Certainly no one could have predicted what has happened. Back then, after 121 others had turned this book down, one lone editor offered a standard $3,000 advance. He said the book forced him to decide what he was in publishing for, and added that although this was almost certainly the last payment, I shouldn't be discouraged. Money wasn't the point with a book like this.

That was true. But then came publication day, astonishing reviews, best-seller status, magazine interviews, radio and TV interviews, movie offers, foreign publications, endless offers to speak, and fan mail—week after week, month after month. The letters have been full of questions: Why? How did this happen? What is missing here? What *was* your motive? There's a sort of frustrated tone. They know there's more to this book than meets the eye. They want to hear all.

There really hasn't been any "all" to tell. There were no deep manipulative ulterior motives. Writing it seemed to have higher quality than not writing it, that was all. But as time recedes ahead and the perspective surrounding the book grows larger, a somewhat more detailed answer becomes possible.

There is a Swedish word, *kulturbärer,* which can be translated as "culture-bearer" but still doesn't mean much. It's not a concept that has much American use, although it should have.

A culture-bearing book, like a mule, bears the culture on its back. No one should sit down to write one deliberately. Culture-bearing books occur almost accidentally, like a sudden change in the stock market. There are books of high quality that are an important *part* of the culture, but that is not the same. They *are* a part of it. They aren't carrying it anywhere. They may talk about insanity sympathetically, for example, because that's the standard cultural attitude. But they don't carry any suggestion that insanity might be something other than sickness or degeneracy.

Culture-bearing books challenge cultural value assumptions and often do so at a time when the culture is changing in favor of their challenge. The books are not necessarily of high quality. *Uncle Tom's Cabin* was no literary masterpiece but it was a culture-bearing book. It came at a time when the entire culture was about to reject slavery. People seized upon it as a portrayal of their own new values and it became an overwhelming success.

The success of *Zen and the Art of Motorcycle Maintenance* seems the result of this culture-bearing phenomenon. The involuntary shock treatment described here is against the law today. It is a violation of human liberty. The culture has changed.

The book also appeared at a time of cultural upheaval on the matter of material success. Hippies were having none of it. Conservatives were baffled. Material success was the American dream. Millions of European peasants had longed for it all their lives and come to America to find it—a world in which they and their descendants would at last have enough. Now their spoiled descendants were throwing that whole dream in their faces, saying it wasn't any good. What did they want?

The hippies had in mind something that they wanted, and were calling it "freedom," but in the final analysis "freedom" is a purely negative goal. It just says something is bad. Hippies weren't really offering any alternatives other than colorful short-term ones, and some of these were looking more and more like pure degeneracy. Degeneracy can be fun but it's hard to keep up as a serious lifetime occupation.

This book offers another, more serious alternative to material success. It's not so much an alternative as an expansion of the meaning of "success" to something larger than just getting a good job and staying out of trouble. And also something larger than mere freedom. It gives a positive goal to work toward that does not confine. That is the main reason for the book's success, I think. The whole culture happened to be looking for exactly what this book has to offer. That is the sense in which it is a culture-bearer.

The receding Ancient Greek perspective of the past ten years has a very dark side: Chris is dead.

He was murdered. At about 8:00 P.M. on Saturday, November 17, 1979, in San Francisco, he left the Zen Center, where he was a student, to visit a friend's house a block away on Haight Street.

According to witnesses, a car stopped on the street beside him and two men, black, jumped out. One came from behind him so that Chris couldn't escape, and grabbed his arms. The one in front of him emptied his pockets and found nothing and became angry. He threatened Chris with a large kitchen knife. Chris said something which the witnesses could not hear. His assailant became angrier. Chris then said something that made him even more furious. He jammed the knife into Chris's chest. Then the two jumped into their car and left.

Chris leaned for a time on a parked car, trying to keep from collapsing. After a time he staggered across the street to a lamp at the corner of Haight and Octavia.

Then, with his right lung filled with blood from a severed pulmonary artery, he fell to the sidewalk and died.

I go on living, more from force of habit than anything else. At his funeral we learned that he had bought a ticket that morning for England, where my second wife and I lived aboard a sailboat. Then a letter from him arrived which said, strangely, "I never thought I would ever live to see my 23rd birthday." His twenty-third birthday would have been in two weeks.

After his funeral we packed all his things, including a secondhand motorcycle he had just bought, into an old pickup truck and headed back across some of the western mountain and desert roads described in this book. At this time of year the mountain forests and prairies were snow-covered and alone and beautiful. By the time we reached his grandfather's house in Minnesota we were feeling more peaceful. There in his grandfather's attic, his things are still stored.

I tend to become taken with philosophic questions, going over them and over them and over them again in loops that go round and round and round until they either produce an answer or become so repetitively locked on they become psychiatrically dangerous, and now the question became obsessive: "Where did he go?"

Where did Chris go? He had bought an airplane ticket that morning. He had a bank account, drawers full of clothes, and shelves full of books. He was a real, live person, occupying time and space on this planet, and now suddenly where was he gone to? Did he go up the stack at the crematorium? Was he in the little box of bones they handed back? Was he strumming a harp of gold on some overhead cloud? None of these answers made any sense.

It had to be asked: What was it I was so attached to? Is it just something in the imagination? When you have done time in a mental hospital, that is never a trivial question. If he wasn't just imaginary, then were did he go? Do real things just disappear like that? If they do, then the conservation laws of physics are in trouble. But if we stay with the laws of physics, then the Chris that disappeared was unreal. Round and round and round. He used to run off like that just to make me mad. Sooner or later he would always appear, but where would he appear now? After all, really, where did he go?

The loops eventually stopped at the realization that before it could be asked "Where did he go?" it must be asked "What is the 'he' that is gone?" There is an old cultural habit of thinking of people as primarily something material, as flesh and blood. As long as this idea held, there was no solution. The oxides of Chris's flesh and blood *did,* of course, go up the stack at the crematorium. But they weren't Chris.

What had to be seen was that the Chris I missed so badly was not an object but a pattern, and that although the pattern included the flesh and blood of Chris, that was not all there was to it. The pattern was larger than Chris and myself, and related us in ways that neither of us understood completely and neither of us was in complete control of.

Now Chris's body, which was a part of that larger pattern, was gone. But the larger pattern remained. A huge hole had been torn out of the center of it, and that was what caused all the heartache. The pattern was looking for something to attach to and couldn't find anything. That's probably why grieving people feel such attachment to cemetery headstones and any material property or representation of the deceased. The pattern is trying to hang on to its own existence by finding some new material thing to center itself upon.

Some time later it became clearer that these thoughts were something very close to statements found in many "primitive" cultures. If you take that part of the pattern that is not the flesh and bones of Chris and call it the "spirit" of Chris or the "ghost" of Chris, then you can say without further translation that the spirit or ghost of Chris is looking for a new body to enter. When we hear accounts of "primitives" talking this way, we dismiss them as superstition because we interpret *ghost* or *spirit* as some sort of material ectoplasm, when in fact they may not mean any such thing at all.

In any event, it was not many months later that my wife conceived, unexpectedly. After careful discussion we decided it was not something that should continue. I'm in my fifties. I didn't want to go through any more child-raising experiences. I'd seen enough. So we came to our conclusion and made the necessary medical appointment.

Then something very strange happened. I'll never forget it. As we went over the whole decision in detail one last time, there was a kind of dissociation, as though my wife started to recede while we sat there talking. We were looking at each other, talking normally, but it was like those photographs of a rocket just after launching where you see two stages start to separate from each other in space. You think you're together and then suddenly you see that you're not together anymore.

I said, "Wait. Stop. Something's wrong." What it was, was unknown, but it was intense and I didn't want it to continue. It was a really frightening thing, which has since become clearer. It was the larger pattern of Chris, making itself known at last. We reversed our decision, and now realize what a catastrophe it would have been for us if we hadn't.

So I guess you could say, in this primitive way of looking at things, that Chris got his airplane ticket after all. This time he's a little girl named Nell and our life is back in perspective again. The hole in the pattern is being mended. A thousand memories of Chris will always be at hand, of course, but not a destructive clinging to some material entity that can never be here again. We're in Sweden now, the home of my mother's ancestors, and I'm working on a second book which is a sequel to this one.

Nell teaches aspects of parenthood never understood before. If she cries or makes a mess or decides to be contrary (and these are relatively rare), it doesn't bother. There is always Chris's silence to compare it to. What is seen now so much more clearly is that although the names keep changing and the bodies keep changing, the larger pattern that holds us all together goes on and on. In terms of this larger pattern the lines at the end of this book still stand. We *have* won it. Things *are* better now. You can sort of tell these things.

(This last line is by Nell. She reached around the corner of the machine and banged on the keys and then watched with the same gleam Chris used to have. If the editors preserve it, it will be her first published work.)

—ROBERT M. PIRSIG
Gothenburg, Sweden
1984